AF352650

The Ephemeral History of Perfume

The Ephemeral History of Perfume

Scent and Sense in Early Modern England

HOLLY DUGAN

The Johns Hopkins University Press

Baltimore

The Johns Hopkins University Press
2715 North Charles Street
Baltimore, Maryland 21218-4363
www.press.jhu.edu

Library of Congress Cataloging-in-Publication Data

Dugan, Holly, 1975–
The ephemeral history of perfume : scent and sense in early modern England /
Holly Dugan.
p. cm.
Includes bibliographical references and index.
ISBN-13: 978-1-4214-0234-5 (hardcover : alk. paper)
ISBN-10: 1-4214-0234-3 (hardcover : alk. paper)
1. Smell—History—Sources.   2. Perfumes—England—History.
3. England—Social life and customs—17th century.   I. Title.
BF271.D84 2011
152.1'660942—dc22   2011004058

A catalog record for this book is available from the British Library.

Special discounts are available for bulk purchases of this book. For more information, please contact Special Sales at 410-516-6936 or specialsales@press.jhu.edu.

For my family, especially my father

CONTENTS

ACKNOWLEDGMENTS

I have worked on this project for almost a decade. It thus bears an unflinching trace of my development as a scholar and of time spent in Ann Arbor, in New York, and in Washington, D.C. At the risk of sounding maudlin, the debt of gratitude I have amassed in each of these places far exceeds what I can acknowledge here. There are many who I simply cannot thank enough, but I will try. Anne Herrmann, Carol Karlsen, Peggy McCracken, Susan Scott Parrish and the amazing Valerie Traub believed in this project before I did. I am grateful for their patience in its early stages and their incisiveness in its final stages as a dissertation, especially Valerie, whose mentorship far exceeded anything I could have hoped for from a dissertation chair. Her tenacity, political conviction, and scholarly investment in both her work and mine have influenced how I define success in this field.

To say that Ann Arbor winters are bleak is an understatement; that I am sometimes nostalgic for them speaks to the warmth of camaraderie I found in my peers, both in Women's Studies and in English. Four friends in particular made my time there (and so many other places) so much fun: many, many thanks to Ezequiel Berdichevsky, Erika Gasser, Nicholas Syrett, and Kelly Williams. I doubt I would have made it through graduate school without them. Ezequiel Berdichevsky, and his lovely wife Aarti, introduced me to Paul, and for that, I am forever grateful. Erika Gasser is the funniest person I know, and perhaps also the kindest. Nicholas Syrett generously shared his rent-controlled Manhattan apartment with me, along with his snarky good sense. And Kelly Williams, from the first moment of graduate school, has been the very best of friends. She has read every line of this book more than twice without ever tiring of it. Her influence is on every page, though the mistakes are all mine.

I am especially grateful to my amazing colleagues at the George Washington University and in Washington, D.C., who inspire me with their dazzling range of expertise, especially Masha Belenky, Leah Chang, Jeffrey Jerome Cohen, William Cohen, Patrick Cook, Maria Frawley, Thomas Guglielmo, Jennifer James, Jonathan Hsy, Constance Kibler, Antonio Lòpez, Rachel McLaughlin, Madhavi Menon, Robert McRuer, Marcy Norton, Samantha Pinto, Linda Salamon, Linda Terry, Sergio Waisman, Gayle Wald, Tara Wallace, Sarah Werner, Lynn Westwater, and the amazing Jonathan Gil Harris. I have learned so much from them. Two groups, in particular, have shaped my thinking about the history of the body: GWU's Medieval and Early Modern Institute and the DC Queer Theory reading group. Both believe heartily that rigorous, academic debate goes best with booze and food. They are my kind of scholars.

My friends Alan Francisco-Tipgos, Joseph Tully, Robert Nakatani, Rose Saxe, Mary Courtney McKee, and Mahmoud Hacheni deserve special thanks since they, among so many other things, provided me with much needed breaks. Time spent with them in Brooklyn, New York; Mattituck, Long Island; Paris, France; and Barnegat Light, New Jersey (yes, the Jersey shore!) allowed me to relax and enjoy those summers when so much writing occurred. Likewise, my amazing 'band of brothers' in medieval and early modern studies, usually seen in short bursts at conferences, equally inspired me to keep writing, shaping how I approach this topic through discussions over beer, coffee, and many, many SAA seminar tables. Thanks especially to Sabiha Ahmed, Laura Williamson Ambrose, Amanda Bailey, Gina Bloom, Carol Brobeck, Joseph Campana, Marlene Lynette Eberhart, Jennie Evenson, Lara Farina, Will Fisher, Roze Hentschell, Derrick Higginbotham, Emily Isaacson, Miriam Jacobson, Melissa Jones, Joseph Lowenstein, Scott Maisano, Lucy Munro, Barbara Sebek, Adam Smyth, and Andrew Tumminia. My students at the George Washington University thought through these ideas with me, offering their own dazzling insights, especially Elizabeth Blake, Ashley Denham Busse, Tessa Kostelc, Nedda Mehdizadeh, and Jennifer Wood along with the members of my freshmen dean's seminar on the boundaries of the body and my graduate seminar on the intimate senses of taste, touch, and smell.

Most of all, I'm grateful for my family. Paul LaMade is the very best husband, father, and friend; without his support, I would not have finished this book. Our son, Brian, was born just as I was finishing this manuscript. Watching him discover his sensory world is among the greatest joys I have ever known. That I get to share this joy with Paul is beyond the beyond. His faith in me—as a mother and as a scholar—means more to me than I can ever express. I am so thankful for my amazing siblings—my brother-in-law, Jonathan; my brother, Christopher; his wife, Tara; and my sister, Melissa—and for my parents, Theadosia and Eugene. I dedicate this book to all of them, and especially, to my dad. For almost as long as I have been working on it, I have joked that he is the only person who will ever read it. Last year, when we almost lost him, I realized how much that meant to me. I am so, so grateful that I have the chance to share it with him.

I could not have pursued this project without generous travel grants, and fellowship support, from the University of Michigan's Horace H. Rackham Graduate School, Institute for Research on Women and Gender, Center for European Studies, Department of English, and Women's Studies Program, as well as from The George Washington University's Office of the Vice President of Research and its Department of English, the Folger Shakespeare Library, and the Huntington Library. My evidence would be scant without the help of archivists in the the Beinecke Library, British Library, the Clements Library, the Folger Shakespeare Library, the Guildhall Library, the Huntington Library, the New York Public Library, the Wellcome Library, the Worshipful Society of Apothecarists' Records Hall, the National Museum of Science, the Victoria & Albert Museum, the Metropolitan Museum of Art, the Cloisters, the Museum of Costume History in Bath, and the Osmothèque, especially M. Jean Kerleo, perfumer, co-founder, and director of the Osmothèque, and Thomas Dungen, an independent scholar who generously shared his research on London perfumers with me. I thank the Marquess of Salisbury for permission to quote from his manuscripts, and his archivist, Robin Harcourt Williams for his help with them. Portions of the introduction, chapter 1, and chapter 5 have appeared in *Literature Compass,* the *Journal of Early Modern Culture,* and *Working Subjects in Early Modern England,* ed. Michelle Dowd and Natasha Korda (Aldershot, U.K.: Ashgate, 2011). Thanks to Michael Cornett, Michelle Dowd, Natasha Korda, Lucy Munro, and the anonymous reviewers of those pieces for their editorial guidance, as well as Blackwell Publishing, Duke University Press, and Ashgate Press for permission to republish these portions here. I would also like to thank Bob Brugger, Matthew McAdam, and Deborah Bors, my editors at the Johns Hopkins University Press, and Glenn Perkins, copyeditor extraordinaire, for their help. Finally, the anonymous reviewer of this manuscript offered generous and incisive comments in a timely manner, which, to a nervous, pre-tenure faculty member, meant everything. The professional courtesy he or she extended to me cannot be repaid directly, but I promise to pay it forward.

The Ephemeral History of Perfume

Strange, Invisible Perfumes

> But when from a long-distant past nothing subsists, after the people are dead, after the things are broken and scattered, taste and smell alone, more fragile but more enduring, more unsubstantial, more persistent, more faithful, remain poised a long time, like souls, remembering, waiting, hoping, amid the ruins of all the rest; and bear unflinchingly, in the tiny and almost impalpable drop of their essence, the vast structure of recollection.
>
> Marcel Proust, *Remembrance of Things Past*

The mysterious power of smell to evoke a "remembrance of things past" (or what scientists now term a "Proustian" or "involuntary" memory) was recently revealed. On October 4, 2004, Dr. Richard Axel and Dr. Linda B. Buck were awarded the Nobel Prize in Physiology or Medicine for their groundbreaking discoveries of "odorant receptors and the organization of the olfactory system."[1] The *New York Times* framed their discovery in more general terms: the science section boldly proclaimed that Axel and Buck had won the Nobel for "demystifying the sense of smell," the "most enigmatic sense," by clarifying how the human body "can recognize and remember" more than ten thousand distinct odors.[2] In announcing the award, the jury noted, like Proust, that smell was essential both to survival and to the enjoyment of life's more subtle pleasures, like enjoying a good wine or appreciating a beautiful lilac.[3] Solving a two-thousand-year-old scientific puzzle, Axel and Buck had cracked this intimate, enigmatic, and ephemeral sense, mapping Proustian memory through a "molecular logic" of smell.[4]

In the press that followed this announcement, however, the metaphoric logic of smell remained enigmatic, odd, and at times, gross. The *Times* immediately ran a secondary piece on the award, noting that it carried "a whiff of spoof science" and linking it to the previous month's Ig Nobel prize, which honored a "bizarre" biological breakthrough in herring communication.[5] Herring appear to communicate through the release of bubbles, or by "farting."[6] The researchers of the fish study, however, warned that any connection to Axel and Buck's work was minimal since the "fish flatulence" in question were not digestive gasses and, thus, carried no odors.[7] In a similar vein, the *Wall Street Journal* ran a front-page piece the following week on J. M. Knight, creator of "odd" and "evil" scents for a variety of educational and consumer applications, including those that mimicked the smell of locker rooms, meteorites, mummies, body odor, dragon's breath, machinery, rainforests, "evil" cotton candy, rotten eggs, Havana cigars, iron smelting, "granny's kitchen," coal fire, flatulence, and, perhaps most troubling, a "Japanese prisoner of war."[8] Both headlines punned that the smell of success is not always sweet.

As both the Nobel and Ig Nobel awards demonstrate, smell bridges acute sensory perception and brute bodily materiality. Both exist within a specific spatial and temporal setting. Axel described his work on olfaction as a study of "the astonishing problem of how the brain represents the outside world . . . how it is that the rich array of mechanical, optical, and chemical properties that define touch, hearing, vision, smell, and taste can be represented by bits of electrical activity that can essentially only vary in two parameters: time and space."[9] Invisible chemical properties of smell are inhaled by the nose and then are translated into electrical charges that travel throughout the brain. These charges form patterns, or smell memories, that can be recalled at a later point in time (if experienced again). Axel and Buck's work demonstrates how, once smelled, a rose is a rose is a rose (at least in terms of brain recognition).[10] Yet, as Axel's meditation on perception also suggests, the representation of that rose (whether as electrical energy or as a sonnet) depends on particular encounters in time and space. Though the human body is programmed to instantaneously recognize more than ten thousand different scents, the cultural and material meanings of those scents vary greatly.[11]

As the 2004 Nobel Prize in Physiology or Medicine recognized, smell is culturally and biologically central to human life, yet it often seems enigmatic. Though scientists have unlocked the molecular logic of olfaction, its role in the past remains a mystery, virtually ignored in historical scholarship. This book explores how time and space determine the metaphoric and material history of smell, arguing that scents are cultural materials worthy of historical investigation. Taking early modern English perfume as my subject matter, I argue that its ephemeral history provides a unique opportunity to examine historical relationships among materiality, perception, and representation while challenging implicit assumptions about the universality of sensory perception and the history of the human body.

Early modern English scents are ephemeral in two key ways: as material objects and as objects of historical investigation. Early modern smells, and any meanings they once held, faded long ago. Likewise, their impact on cultural histories of early modern England is almost nonexistent. Thus, to those familiar with popular histories of perfume or with the social and cultural histories of early modern England, writing about the intersection of the two may seem surprising. Accounts of the history of perfume uniformly assert that modern perfuming practices originated in ancient Egypt, were elevated to an art-form in ancient Rome, lost in medieval and Renaissance Europe, and resurrected in eighteenth-century France with the birth of modern perfuming industries. Within such trajectories, early modern English scents hardly seem relevant at all.

In the hope of dis-*orient*-ing this exoticized history of perfume, I begin my exploration of the relationship between the history of the senses and the history of scents in an unlikely place: early modern England. Early modern England rarely factors into "natural histories" of perfume; if mentioned at all, it is to provide an interesting perfuming anecdote—perhaps Henry VIII's voracious appetite for scented waters or Elizabeth I's queer preference for elaborately scented leather gloves. Although early modern England is often imagined as a smelly place, it is not imagined as a cultural environment in which *perfume* factored largely. Despite participating in numerous scientific and colonial endeavors that defined what we now call the Renaissance, England remains an undiscovered country within the history of olfaction. Why does the Renaissance factor so blankly in popular mythologies about perfume's history? And why is England absent from modern cartographies of perfume? As the following chapters will demonstrate, the obvious answer—that it was not culturally or materially significant—is simply wrong.

Likewise, my subject matter may seem equally surprising to those familiar with early modern English culture. Perfume does not figure largely in most studies of early modern England, a period in which new smells and new cultural theories about olfaction were developed (particularly concerning perfume). Indeed, many of the anecdotes of early modern scent recounted in my chapters were relegated to the footnotes of other scholarly works. Others were found in odd or unlikely places: medieval hagiographies, guild records, medical texts, new world contact narratives, plague accounts, plays, and gardening manuals. Though sensory history, or what Bruce R. Smith has termed "historical phenomenology," has emerged as a viable field of scholarly inquiry, there have been few in-depth studies of olfaction.[12] And although there have been important scholarly studies of the history of hearing, touch, and taste (in several historical periods, including early modern England), very little scholarship has been done on premodern histories of olfaction.[13]

This lack of scholarly attention stems partly from an erroneous belief about the relationship between the material history of smell and metaphoric language. As C. M. Woolgar argues in his study of the senses in late medieval England, "Smell is unlike other senses in that we lack a specific vocabulary to describe many of the sensations that can be perceived through this faculty. Some can only be put into words by borrowing terms descriptive of other senses, particularly taste, for example sweet, bitter or acrid; or, by analogy, things 'smell of something,' such as roses or decay."[14] For Woolgar, this problem is compounded by a dearth of material evidence; because material smells are ephemeral, "we are entirely dependent on written descriptions for our information about the sense and its operation."[15] Whereas Woolgar asserts that this dearth of evidence was true of the late medieval period, I disagree with his assessment of early modern material and metaphoric histories of olfaction. Early modern English had a precise language of olfaction that described the powerful and invisible interaction between scents and people.

Nor is metaphor necessarily a hindrance to historicizing smell. The relationship between a physical sensation, like the sweet smell of an actual rose, and a cultural representation of it, like its use in the play *Romeo and Juliet,* is a gap filled by metaphoric language. That even biologists describe the instantaneous nasal recognition of a smell molecule as a "Proustian" memory emphasizes the important role of metaphor in both biological and cultural experiences of sensation. Modern, international scientific discourses understand "smells" to be transmitted in airy molecular forms: we "smell" a molecule when it is inhaled and hits one of our genetically programmed smell receptors in the nose. As such, gas molecules are material objects, but until recently, these molecules were only identifiable through language descriptions. These invisible objects—smells—have been interpreted as gassy molecules, chemicals, contagions, even bodily emissions, and, as I will argue, as spice, luxury, flowers, musk, the presence of God, ethnic others, disease, sex, or gender; we need to attend to the role that metaphor plays in relationship to the material history of smell.

Metaphor is the apparatus through which invisible smells are "made to appear."[16] George Lakoff and Mark Johnson argue that metaphoric language betrays cognitive patterns of sensation: "metaphor is as much a part of our functioning as our sense of touch, and as precious."[17] As these scholars suggest, our subjective archive of sensory experience is shaped by culturally shared metaphoric tags, like "rose" or "sweet." The study of metaphor thus can "make available to others the experience of the corporeal senses."[18] Metaphors can function as a historical archive of sensation. This is not to say that metaphors should be interpreted literally or that they render themselves meaningful only through shared cultural histories.[19] Rather, they reveal how individuals react to cultural and physical events. If interpreted thoughtfully, metaphors demonstrate the potential of literary history to offer "abstract" and "concrete" contributions to other fields like quantitative history, geography, and botany.[20]

This is especially true of early modern metaphors of sensation; early modern English included a much larger vocabulary of sensory description than most scholars assume, a vocabulary that is now mostly obsolete.[21] Although Renaissance science lacked any cohesive understanding of how such reception occurred, believing that the nose, the mouth, and even the womb could "smell," early modern men and women had distinct ways to speak about perfume and its effects, with words that described the effects of perfume, incense, and scent in religious, medicinal, and sexual experiences. Objects ambered, civited, expired, fetored, halited, resented, and smeeked; they were described as breathful, embathed, endulced, gracious, halited, incensial, odorant, pulvil, redolent, and suffite. Scent descriptions included marechal (cherry), naphe (orange), thymiama (incense), and suffiments (general terms for medicinal scents), and they existed as diaspasms, powders, sainses, smokes, waters and balms. In the sources surveyed in this book, Renaissance commentators on olfaction (including scientists, artists, guild apprentices, playwrights, housewives, and theologians) were troubled more by the invisible nature of scent than they were by linguistic imprecision.

The ephemerality of the history of early modern English perfumes thus seems paradoxical. There is a profound insistence that the past was smellier than the present. It is often casually pointed out to me that people in the past *must* have smelled bad (or, at least, strong) to others in the past (as opposed to now). Such comments reveal that many of us are willing to believe that smell mattered in the past; however, we remain unwilling to acknowledge that it might produce or represent embodied physicality or matrixes of social relationships in the present As feminist and postcolonial historians, literary critics, and psychologists have noted, assertions about the universal appeal of deodorized bodies reveal the sexist and racist coding of smells in our own culture.[22] Lost in historical abstraction, smell becomes abstract and universal—an unconscious emission, a fact of biology, a part of a primitive past rather than a civilized present—in short, representative of a materiality located outside the self that obscures the particularities of present and past struggles over embodiment. Our cultural willingness to belie the material history of smell may have more to do with present concerns over the social meanings of embodiment than with archival obstacles.

Within some studies of the senses, the cultural importance of scents in the past depends on assumptions about their denigration in the present: smelling, we assume, must have been important in pre-Enlightenment culture since, without modern hygiene, the world of the past must have been smellier than our own. Early work on the senses promoted this theory of the "great divide" between premodern and modern culture: Renaissance technologies of print were critical in elevating sight (and the eye) to the domain of reason and objectivity, leading to Enlightenment scientific practices.[23] Sensory hierarchies of modernity elevated sight above all others, relegating the lower orders of smell, taste, and touch to the primitive past.[24]

Most cultural histories of the senses thus chart only olfaction's demise in post-Enlightenment Western cultures.[25] In such scholarship, sorting out the subjects and objects of olfaction emerges as a civilizing process: modern Western bodies become subjects who smell when they stop being objects that smell. Buoyed by this teleology, most cultural and "natural" histories of scent invoke a methodology of inversion: utilizing olfaction's lack of historicity as a tool to breach smells of the past, the authors advise paying more attention to smells in the present. In an effort to undo the effects of western "civilized" culture, a new embodied "enlightenment" emerges: scents become an exotic tool of transformation to another time and place. Yet, such beliefs reflect only modern assumptions about the sensations of everyday life in the past.

New Sensations: Archives of Olfaction

Once the material and metaphoric history of perfume is placed firmly within canonical understandings of the English Renaissance, it becomes clear, as Proust and scientists agree, that olfaction does have a history and that it comprises a rich archive of everyday life in the past. Interpreting this archive, however, presents logistical and methodological problems. First and foremost, smells are invisible, which challenges implicit assumptions about their materiality. In his landmark study of phenomenology, Edmund Husserl argued that objects are things that can be handled, displayed, and—most importantly—seen: "If one speaks simply of objects, one normally means actual, truly existing objects . . . no matter what one says about such objects, that which is meant and stated must—if one speaks rationally—be something which can be 'grounded,' 'shown,' directly 'seen.'"[26] To speak of invisible scents as material objects risks irrationality. In this book, however, I question the limitations of Husserl's claim and argue that visibility and materiality were not always historically linked. In doing so, I contend that many objects were valuable in the Renaissance precisely because *they were not meant to be seen;* they were meant to be touched, tasted, heard—or as I will demonstrate—smelled.

All of the bodily senses participated in the creation, description, and distribution of material objects and defined the limits of what could be perceived, manipulated, and represented in early modern culture.[27] I am not alone in asserting this claim; its implications certainly extend beyond the early modern period.[28] In his summary of the field, "Producing Sense, Consuming Sense, Making Sense: Perils and Prospects of Sensory History," Mark Smith asserts that it is a good moment to be a sensory historian.[29] Sensory history is a burgeoning field; what was once described as a "senseless" profession is now rife with sensuous explorations and encounters with the past.[30] Recent scholarship on the early modern period includes studies of the acoustic world of the Renaissance, of the role of voice in creating gender on the Renaissance stage, of the role of touch in early modern culture, of scents as staged properties, of taste in early modern manuscript coteries, of chocolate and tobacco in early modern Europe, and of the sensory worlds of early America, to name just a few examples.[31] Though these studies are diverse in their approaches and arguments, collectively they demonstrate that the five human senses—vision, hearing, smell, taste, and touch—provide an important way for scholars to understand the interface between material environments and somatic experiences of the past.

Until recently, such a conclusion seemed oxymoronic, if not impossible: how could something as subjective, fleeting, and ephemeral as the sensate have a history? Though Aristotle first defined the five human senses over two thousand years ago, definitions and theories about them have varied widely. Nor has there been agreement that the human senses are limited to five; for example, Augustine posited that there was a sixth, inner sense, which perceived not only external objects but also the five senses themselves. Contemporary scientific discourse now recognizes nine.[32] Belief in five, six, or nine senses raises important questions about their role as historical evidence: how might we explain such differences? Has the human body adapted and developed new modes of perception or merely new theories about them?

Mired in misconceptions about biology, cognition, and representation, the senses represent a provocative threshold of interpretation and a historical paradox: is the body a static or shifting category of knowledge? As literary critic

Michael Schoenfeldt argues, such differences result from cultural beliefs: "bodies have changed little through history, even though the theories of their operations vary enormously across time and culture . . . [w]e are all born, we eat, we defecate, we desire and we die."[33] Though this eloquent observation offers a seemingly overarching truth about human experience, it obscures the ways individual bodies sense specific phenomena, transforming both the body and the object being sensed. When one imagines such a body, it is hard not to import assumptions about what does and does not comprise its materiality. For example, does is matter if it is tall, muscular, amputated, wounded, pregnant, or starving? Are these descriptions merely cultural theories of operation or bodily facts?

Put another way, as Thomas Lacquer argues in *Making Sex,* there are differences between "the body, and the body as discursively constituted, between seeing and seeing-as."[34] The stability of the body as both a material and discursive site of meaning simultaneously ensures and obscures the history of olfaction: one can presume that, like birth, reproduction, and death, olfaction existed in the past. In order to eat, defecate, and desire, we must first breathe, and, therefore, smell. The meanings of olfaction, however, vary greatly across time and culture and are equally grounded in a material reality, depending on and changing according to the (imagined) experience of smell. But what if a body produced the smell being investigated? The difference between the discursive and material body is a confusing threshold of meaning if one seeks to chart not the difference between seeing and seeing-as but between smelling and smelling-as. Do all bodies smell the same?

Judith Butler, *pace* Foucault, questions the reflexivity implicit in this question, exploring what she identifies as a Western philosophical insistence on the stability of the material body as a site of sexual (rather than historical) difference. In doing so, she argues that the body as a stable, intelligible site (always four limbs, five senses, or the sex one simply "has") begs the question of the body's materiality and boundaries. What must remain unintelligible, immaterial, and "radically dematerialized" in order for us to understand the somatic fabric of our being?[35] Put another way, in order to understand the cultural and biological meanings of smell in any given historical or cultural moment, one must grapple with the body's material *and immaterial* matter.

Embodiment is the sum of these material differences. It is a biological and discursive cultural phenomenon, an "open-ended category," and a "privileged operator for the transcoding" of social formations, symbolic typographies, and textual metaphors.[36] Embodiment is often imagined as a privileged site through which to explore "the witnessed and perceived particular[s]" of life in the past, emerging as a cultural phenomenon worthy of historical study.[37] Critics have defined *embodiment* as interpretations of a perceiving subject located within somatic materiality.[38] The history of embodiment represents a threshold of understanding about lived experiences of the past. In many ways, embodiment, as a threshold of knowledge, demonstrates the challenges inherent in historicizing olfaction. Variously described as "the interface between the psychic and the corporeal," "materialist psychology," "embodied understanding," "humoral materialism," the "psychophysical self" or "differences between within and without," embodiment is critically defined through attempts to fuse premodern and postmodern approaches to the somatic material of bodies and the internal realm of minds.[39]

Embodiment, as a critical term, potentially links disparate work on the body, including historiographic models that define *body* as a material, social, and discursive phenomena. Despite what historian Kathleen Canning has termed "a veritable flood" of historical investigations of the body since the 1980s, the term *body* remains "a largely unexplicated and undertheorised *historical* concept," referring simultaneously to critical work on cultural assumptions that define bodily prohibitions and to local histories of individual bodies, situated (like Axel and Buck's smell molecules) within definable locations in time and space.[40] The bodies analyzed within "body history" are either "singularly discursive or abstract," "excessively material and undertheorised," or simply "not made visible at all."[41] Embodiment, for Canning, emerges as a useful concept and method: its critical history seeks to locate various theoretical, material, and invisible bodies within distinct historical social contexts, as respondents to the interplay of heterogeneous cultural spaces.[42]

Fueled by postmodern critiques of objectivity, embodiment is no longer an obstacle to achieving transcendental objectivity; rather, it reflects the endless situatedness of perspective.[43] Consider, for example, historian Peter Hoffer's exploration of the sensory worlds of early America. Hoffer carefully notes that engaging the senses as historical evidence poses obvious limitations: "If in the past our predecessors' use of the senses was culturally bound, how can [historians], bound as well by our culture, recover what others long ago saw, heard, and smelled?"[44] Yet Hoffer suggests that the historian's own body can be a methodological tool for exploring embodiment; living museums, historical reenactments, and traditional archival evidence can train historians to see, hear, taste, touch, and smell in new (and thus old) ways.[45] Embodiment, he argues, represents a threshold of knowledge through which the past might be experienced in the present. Yoking postmodern critiques of objectivity with historical understandings of subjectivity, Hoffer suggests that the historian's own senses contain gateways to the past.

Hoffer's methodology raises serious questions about the concordance between bodies in the past and bodies in the present. It posits that although sensations of the past are radically different from those of the present, historians can—and should—trust their own imagination of what sensation was like in situ. This approach has both problems and tremendous potential.[46] I differ with Hoffer's claim that the historian's body provides its own threshold of meaning through which one might encounter sensory worlds of the past. The differences between sensations of the present and the past are complicated enough; for example, what is commonly known as "musk" to modern perfume consumers bears almost no resemblance to the smell of musk familiar to early modern ones.[47] Merely sniffing musk in the present affords no greater insight into the meanings of musk in the past—it is a materially different smell. There may also be differences in biological mechanisms of smell. Gender and age both affect an individual's ability to perceive smells, and scientists note that modern, environmental factors play a key part in defining one's susceptibility to anosmia, a condition in which one's ability to smell declines over time.[48] Such research suggests how modern olfaction may biologically differ from olfaction in the past.

Yet Hoffer's approach has produced brilliant readings of the sensory worlds of early America. He outlines distinct research opportunities—moments when grappling with sensory worlds of the past reveal the fault lines of conflict and contestation between different groups and their worldviews.[49] How does one deal with new sensations, never encountered before? How does one represent invisible forces within the natural and social world? How do the senses shape contact with such forces and with those perceived as others within it? And how does cultural etiquette shape sensation, that which is experienced firsthand as true?

Grappling with such questions provides new ways of understanding large-scale shifts in cultural approaches to the body. It might also provide a more nuanced way of attending to bodily matter, to the material, discursive, and spatial relationships between bodies and their environs, provided that one carefully situates both the act of sniffing and the objects sniffed in the past in order to understand their cultural meanings. Rather than using the historian's body as a tool in the archive to access sensations of the past, historian Mark Smith suggests that scholars should cultivate an approach to sensory history that posits instead a "way of becoming attuned to the wealth of sensory evidence embedded in any number of texts, evidence that is overwhelmingly apparent once, and ironically, looked for."[50]

There are substantial archives that document how early modern men and women produced, consumed and represented scents and their ephemeral effects. These archives about scent-making provide a rare—and underexplored—opportunity to study how early modern men and women defined and negotiated the environment of their everyday lives. They also challenge assumptions about what comprises historical evidence and materiality itself. For example, early modern scents were not only invisible; they were also ephemeral. Unlike the material traces of early modern everyday life on display in museums or cataloged in archives—clothing, crockery, armor, furniture, or print and manuscript texts—early modern scents no longer exist. Yet unlike other objects that decayed, were destroyed, or lost due to violence or neglect, early modern scents were designed to have a fleeting materiality.

Early Modern Theories of Olfaction

These fleeting scents, however, were powerful tools of transformation through which early modern English men and women negotiated the atmosphere of their everyday lives. Though external to the body, scents could alter its physiognomy. Olfaction blurred distinctions between bodily boundaries and environments. Early modern theories of olfaction attended to this fact, drawing heavily on classical approaches to the senses, which sought to discern how the body smelled by examining what it smelled.[51] Four, in particular, were especially influential on Renaissance theories of olfaction: Plato, Aristotle, Theophrastus, and Galen. Plato argued that there was no clear pattern to human olfaction because of the material instability of smells. One only noticed such "half-formed things" if a particular smell was pleasant or foul, making objective classification impossible.[52] Furthermore, a smell's deformed state provided a clue to its nature; smells, he argued, emerged during states of transition (or what we might term elemental shifts between solids, liquids, and gases) such as liquefaction, decomposition, dissolution, or evaporation.[53] As such, they had the power to radically influence the body, refreshing or, more likely (given the deformed state of smell), harming it.

Aristotle disagreed sharply with Plato about the nature of olfaction, arguing that smells were not objects in and of themselves (or even half-formed objects) but rather immaterial forms or "species" emanating from an object. Examining how air, which is normally imperceptible to the nose, became "affected" by such forms (and thus perceivable by human noses), Aristotle agreed that smells were influential, arguing that they could even affect even inanimate objects. Seeking to answer the question "what is the nature of smell?," Aristotle implicitly arrived at an answer to "what is the organ of smell?": breathing was integral to human olfaction. Theophrastus, influenced by Aristotle's theory of olfactory influence, argued that although smells were inherently ambiguous, one could conclude that bad smells originated from bad things and good smells from good things, establishing an important olfactory pattern amplified throughout medieval Europe in early Christian practices.[54] Likewise, Galen took Aristotle's surprising implications about the organ of smell one step further, concluding that the organ of smell was not the nostrils or the nose but rather the brain. As such, smell was unique among the five senses in that it had no secondary organ of perception.

Renaissance anatomists were influenced by these theories of matter and operation: olfaction was thus a paradoxical sense, justly serving as a gateway between higher and lower orders of perception. Helkiah Crooke's *Mikrokosmographia* (1615), for example, locates smell as the third of the five human senses. Situated precariously between the perceptive senses of sight and hearing and the bodily senses of taste and touch, smell mediated between one's body and mind and between one's human and animal instincts.[55] The nose, similarly positioned between the sensory organs of the eyes and mouth, represented olfaction's middling status within sensory hierarchies. But, as Crooke is careful to note, the nose was a deceptive representative of olfaction: even if one's nose was stuffed full of aromatics, one could only perceive odors after they were drawn to the brain by breath.[56] "[T]here is no perception of odours, except when we do inspire," Crooke writes, "for though you fill the Nose full with Muscke or Ambergreese or other odoriferous bodyes; yea though you should annoynt the whole membrame with sweet oyles, yet you shall haue no perception of odours except you draw in the Ayre by inspiration."[57] With every breath one inhaled invisible particles that "touched" the brain; since every sense is a "kind of touching," the organ of smelling could not receive odors "unlesse the perfume of odoriferous things should touch the same."[58] Thus, perfume could alter humoral imbalances, refreshing and correcting "distempers" of the brain.[59] Smells were both immaterial forms that shaped the brain's perception and physical entities that penetrated (and influenced) the body.[60]

Though olfaction's influence on the body was undisputed, Renaissance anatomists and medical practitioners were unclear about the precise interaction between an invisible smell and the body perceiving it. Stephen Batman's 1582 translation of Bartholomeus Anglicus, for example, explains that in order to perceive a smell, one must "work" the "ayre," "for to take lykensse of the aire, that commeth from the thing that shall be smelled, the which lykenesse the aire hath of the same thing that shall bee smelled."[61] Breathing was necessary to smell an object but only because the air acquires a "likeness" to the smell itself. If the air can acquire such a likeness, perhaps the perceiver could as well. Even Robert Burton's succinct discussion of the organ and medium of smell becomes cumbersome when discussing objects smelled: "The Organ is the Nose, or two little hollow peeces of flesh a little aboue it: the Medium the ayre to men, as water to fish: the Obiect, Smell, arising from a mixt Body resolued, which whether it be a quality, fume, or vapor, or exhalation, I will not now dispute, or of their differences, and how they are caused."[62]

The nose thus symbolized olfaction's ambiguous sensory status. Though it was not an organ of smell, it prominently displayed bodily vulnerability to environmental influences and the act of sniffing, particularly as diseases like syphilis and tobacco-caused cancers ravaged early modern noses. Medical descriptions of the nose emphasize this vulnerability through descriptions of its inner workings, emphasizing the nose's feminine and leaky anatomical parts. Small teats or paps descending from the brain were thought to be located at the top of the nose, collecting smells from the air, whereas the nostrils themselves functioned as gutters or sinks in which effluvia from the brain collected and dispelled.[63] Guy de Chauliac's sixteenth-century treatise on surgery argued that the inner part of the nose consisted of two "nosethyrlles," which are "gutters ascendyng unto the bone of the collatory where as are applicate the addicions mamylares of the brayne where as smell is."[64] Likewise, a late sixteenth-century translation of Aristotle's work metaphorically described the nostrils as the "sink" of the brain, a term synonymous with urban sewers.[65]

The notion that the nostrils contained nipple-like mammary glands persisted. By the time Ambrose Paré published his works of surgery in the late sixteenth century, "mamillary projections" was a scientific term of art rather than mere description, linking olfaction with dangerous environmental influences, including eroticism. John Hall's mid-sixteenth-century translation of Lanfranco's *Most Excellent Woorke of Chirurgerie* outlines that at "extreme endes wythin the forheade, at the vpper end of the nosethrilles certeine eminences lyke in forme to the tetes or neples of womens brestes, and these receyue the odoure, smell and sauoure of all thynges, plesante or noysome, throughe the nosethrylles."[66] In Thomas Tomkis's early-seventeenth-century university play *Lingua: or the Combat of the Tongue, and the Five Senses for Superiority*, Olfactus is described as playing both sides: he will "pollitckly leane to neither part / But stands betwixt the camps as at receite."[67] He himself announces "I lay my head between two spongeous pillows, like fair Adonis 'twixt the paps of Venus, / Where I conducting in and out the wind, / Daily examine all the ayre inspir'd."[68]

As this brief survey of early modern English theories of olfaction demonstrates, olfaction's invisible influence on vulnerable bodies was troubling, and theorists struggled to provide their own strategy of dealing with its ephemeral nature. For example, in a 1669 religious meditation titled *A Holy Oyl and a Sweet Perfume,* Sir James Harrington wrestles with olfaction's recalcitrance to historiographic methods, noting two distinct problems: scent lacks proper objects of study, and its effects are transhistorical. Poised between Renaissance theories about the humoral, porous body and Enlightenment science's quantification of matter, Harrington is troubled by the bodily action of smelling. Arguing that each of the other four senses has a clearly delineated holy purpose, Harrington admits that only olfaction remains a mystery to him. He muses that the sense of smell *must* have a (holy) purpose, even though it escapes him and most of the naturalists he quotes.

The difficulty lies in the abstract, invisible, and timeless nature of scent, as well as its susceptibility to sin. He writes, "this sence of *smelling* . . . hath no objects which properly in themselves considered, can be said to be evil, yet hath it (since the Fall) a sinful property and inordinate, which is always ready to assist the pride and inward corruption of the heart."[69] Since all five human senses were divinely created, olfaction must have a holy purpose (despite its ever-ready sinful nature). Yet its invisible nature forces him to rely on imperfect bodily effects as a gauge: "[I]f venomous and malignant smells, and spirations, have demonstrably, and undeniably, a secret and virulent Power, to destroy life and nature . . . [t]hen assuredly, by the rule of contraries, and according to right reason, redolent, fragrant, cordial, and spiritual odours, scents, aires, and smells, do not only refresh and exhilerate, but fortifie, preserve and nourish our life and beings."[70]

Harrington draws distinctions between life-destroying and life-affirming scents because, as he points out, the sense of smell "hath no objects properly in themselves." One can only register a scent's effects through bodily experiences: scents contain secret, virulent power to "destroy life and nature" or to refresh and exhilarate the body. He concludes, rather anxiously, that odors, scents, airs, and smells are not *always* agents of evil. Through assertion of the rule of contraries and a faith in "right reason," Harrington imagines a divine purpose for smell in general and perfume more specifically, even if discernable bodily effects only prove the reverse. Regardless of divine or evil purpose, Harrington assumes that a scent always affects a body in similar ways, solving the problem of the objectless nature of scent with olfaction's transhistoricity. Olfaction's history, according to Harrington, changes little. He is not alone in making this assertion; such conclusions haunt many cultural engagements with olfaction. Perhaps nothing is more implicit in western cultures than the role of smell in the history of embodiment. The notion that smell has no history remains popular and pervasive since we generally assume that there is no evidence to contradict it.

Belief in olfaction's lack of history insists on the difference between subjects and objects, an assumption that asserts a transcendental history of the body, thereby denying cultural ideology, or what Jean Howard eloquently terms "the obviousness of culture, what goes without saying, what is lived as true."[71] In his *Outline of a Theory of Practice,* Pierre Bourdieu analyzes social practices as ongoing processes of embodiment that produce an ever-shifting matrix of "perceptions, appreciations, and actions" that comprise each single, unconscious gesture. Contained in one simple gesture (like sniffing) resides an "implicit pedagogy, capable of instilling a whole cosmology, an ethic, a metaphysic, a political philosophy."[72] These implicit, inherent, or obvious "cosmologies," or, for my purposes, osmologies, define embodiment as a process; absorbing such cultural conditioning defines how a body moves through social space.[73]

Further, the material instability of olfaction (so eloquently described by Harrington) is equally matched by a profound linguistic reflexivity (at least in English). Consider the question "what does it mean to smell?" Subjects who smell can also be objects who smell. Immediately, the odd nature of olfaction erupts in language: is "smell" a transitive or intransitive verb? The grammatical function of "smell" depends on the experience it seeks to describe. One must rely on implicit reading practices in order to understand the linguistic context. The reflexive grammatical nature of smell extends equally into social and material realms, especially if one asks, as I do in this book, what it meant to smell in another historical period. Here, historical and material context matters: where, when, how, and what smelled? Or, to approach the problem from a different vantage: who smelled? Such an inquiry requires grappling with cultural assumptions about the body (and its boundaries), the history of the senses (and sensory perceptions), and the discursive networks of power that frame both.

Recent critical inquiries into the history of the body and the senses do this very task, emerging to contest widespread belief in the stability of the body and its sensory perceptions throughout history. Scholars have begun to trace the implicit cultural pedagogies, ideologies, and cosmologies of social conditioning as defined through the bodily senses in particular historical settings.[74] In such studies, the cultural meanings of sensory perceptions—and the subjects and objects of human sensation—are procured through a range of new approaches to traditional, historical archives.

As these critical methodologies suggest, the history of olfaction (like that of the other senses) exists in the conceptual space between subjects and objects and between theories of embodiment and theories of materiality. These new approaches attempt to proffer the metaphoric and material abstraction that surrounds the history of the senses through interdisciplinary projects, uniting material (and social) historical research with literary, artistic, anthropological, and sociological theories of embodiment. Olfaction thus provides a rare opportunity to understand how early modern English men and women experienced their own bodies and those around them. Smells comprised, literally and metaphorically, the atmosphere surrounding the process of "making and doing the work of bodies—of becoming a body in social space."[75]

Producing, consuming, and representing scents in the Renaissance involved negotiating smelling subjects, scented objects, and the airy boundaries between them. Early modern culture had distinct technological and spatial practices for understanding (and manipulating) the body and its surrounding environment.[76] Defining *perfume* as my object of study, I use the term broadly to include spices, flowers, herbs, animal parts, trees, resins, and other ingredient used to produce artificial scents, smokes, fumes, airs, balms, powders, and liquids in the period. These scents were used for a myriad of cultural purposes in a range of social spaces. I examine six of them: the scent of incense in church, the smell of rosewater at court, the scent of sassafras in the contact zone, the smell of rosemary in the shut-in household, the scent of ambergris in luxury shopping markets, and the smell of jasmine in pleasure gardens. Surveying a wide variety of evidence, I analyze literary texts alongside other texts and objects, including manuscript and printed cookbooks, guild records, herbals, medical anatomies, gardening manuals, censers, casting bottles, prosthetic noses, pomanders, gloves, and potpourri vases. These materials form a history of perfumery in early modern England. They also document the complicated nexus between scent and sense in a host of environments, elucidating how early modern men and women understood the boundaries between their bodies and these spaces.

Each chapter is organized around three interlocking components: a scent ingredient, a material object used to dispense it, and an environment in which it was deployed. Read together, they emphasize important shifts in perfuming practices across the sixteenth and seventeenth centuries in early modern England. The first two chapters examine scent's rarified ability to stage divinity or royalty; the second two examine perfume's role in the medical history of olfaction; the final two examine perfume's emergence as a mass-produced, luxury commodity. Implicit in my argument is an understanding that the production of scent in early modern England was both a technological and artistic endeavor; latent in efforts to make scents are both opportunities and limitations for transgressing shared social networks of spatial practice and altering the body's relationship to its environment. In the alchemy between what could (and could not) be transformed through perfumed scents lie early modern understandings of the role of material culture and artistic representation in describing, affecting, and changing human perception.

In chapter 1, "Censing God," I examine perfume's use as a tool of religious and economic transformation in late medieval England. As recent scholarship has demonstrated, the ritual use of incense in the medieval liturgy transformed prayer, objects, and realms into spaces inhabited by the divine. When new kinds of aromatics entered England's markets in the late fifteenth century, this olfactory marker of divinity began to signify other kinds of transformations, namely England's participation in a global spice trade. Scents like incense were produced from a complicated amalgamation of foreign and native ingredients; these new amalgamations challenged perfume's singular ability to represent divinity even as they increased awareness of the geographic, political, and economic networks that produced such perfumes. The exotic, the profane, and the holy all were invoked by scented incense and balm. No other figure better represents the ambiguous nature of scent in this period (and its complicated relationship to divinity and luxury) than the saint who had it as her emblem: Mary Magdalene. Reading the Digby *Mary Magdalene,* a late-fifteenth-century play, alongside English censers, sermons, and paintings, I argue that her encounter with perfume is representative of a large-scale cultural shift from censing to sensing.

Chapter 2, "Casting Selves," examines how new technologies of distillation affected religious and erotic perfuming practices in the mid-sixteenth century, particularly those involving rose attar. Arguing that the scent of the damask rose was an essential component of both Henry VIII's and Elizabeth I's performances of royal power, I examine how rose perfumes came to define emerging English identity through olfaction. Contrasting early-sixteenth-century Tudor performances with late-sixteenth-century ones, I examine how theatrical space redefined the body and its boundaries. Rose attar, produced from English distilleries and dispensed from casting bottles, became an important part of staging an embodied self.

Chapters 3 and 4 examine perfume's role in medicinal history, examining the link between sassafras and syphilis in the sensory realm of the contact zone and the role of perfume in the shut-in households ravaged by the plague. In chapter 3, "Discovering Sassafras," I examine how perfume shaped Anglo-Indian encounters in the new world. English explorers learned early in the sixteenth century that their success in the new world depended on mastering new strategies of "discovery," including olfaction. Sassafras in particular was critical to English success. Valued for its aromatic properties, especially as a treatment for the pox, sassafras was difficult to locate. English explorers seeking it had only their noses (or native informants' noses) as guides. Discovering sassafras required more than visual mastery of new landscapes; it required an acute awareness of the plant's olfactory properties within a shifting sensory realm. Such practices were not without risk. Sassafras reveals that the phenomenology of discovery (when one switches from technologies of sight to technologies of olfaction) was proximate and dangerous, but potentially rewarding.

In chapter 4, "Smelling Disease," I continue my investigation of the nose as a vulnerable orifice, susceptible to environmental contaminants. In this chapter, I examine the space of the plague-ridden, "shut" household. Because perfume could permeate an enclosed environment, people used it for combating disease, employing rosemary as a plague preventative, for example. Rosemary's abundance and affordability made it a state-sponsored preventative for even the poorest of neighborhoods. Likewise, pomanders stuffed with more expensive spices enabled health practitioners and body-searchers to navigate such dangerous realms. However, perfume's role as a plague preventative caused the sweet scent of perfume to be associated with other, undetectable, airborne contaminants. In this chapter, I examine how perfume came to be seen as an efficacious, environmental tool of debauchery. Analyzing the ability of pleasant odors to "refresh," "exhilarate," and "cheare" a body (along with their ability to tire, weaken, and harm it), I argue that scent was both a tool to fight disease and an aphrodisiac that could inspire contagion.

The final two chapters examine early modern England's economic and commercial history of perfumery. In chapter 5, "Oiled in Ambergris," I examine a pervasive fad for perfumed gloves in "the Spanish style," or gloves heavily scented with ambergris. Ambergris, a product of the sperm whale, had a powerful aroma that could mask the noxious scent of treated leather, enabling the perfuming of gloves. In late sixteenth-century England, a wide variety of guilds, artisans, housewives, and merchants sought to control the lucrative production and consumption of these gloves scented with ambergris. Their economic struggles to control a lucrative ingredient like ambergris created sharp distinctions between medicinal, culinary, pharmacological, fashionable, and cosmetic uses of perfume. By the end of the 1630s, "perfumer" emerged as a recognizable professional identity and perfumes as a recognizable commodity. The history of scented gloves reveals that all sorts of men and women were "oiled in ambergris," isolating the uneasy relationship between production and consumption of perfumes.

Chapter 6, "Bowers of Bliss," analyzes the emergence of botanical scents in commercial perfumes. In the seventeenth century, scents were increasingly categorized into foul or fragrant smells. While foul-smelling odors (like those associated with disease) were regulated under the aegis of public health, fragrant scents and perfumes increasingly became associated with private realms. Historians of hygiene mark a profound shift in the seventeenth century away from heavy, animal-based musk toward fragrant, "natural" scents of botanicals. The perfumes dispensed from English pleasure gardens and potpourri vases exhibited a delicate balance between nature and artifice. Pleasure gardens and potpourri vases were designed to create multisensorial zones of pleasure. Jasmine, in particular, represented a cultivated luxury that scientific advancements in horticulture afforded. An exotic flower grown in the East and West Indies, jasmine was difficult to produce in England; its scent made it a highly prized specimen in private greenhouses and gardens. Potpourri was used to recreate jasmine's cultivated luxury (and the pleasures it inspired) indoors as well as out. Smoke perfumes, potpourri, and other means of scenting airs document how perfumes helped redefine the bedroom as an enclosed, erotic zone like that of the pleasure garden.

Finally, in a brief epilogue titled "Ephemeral Remains," I examine how the ephemeral history of early modern English perfumes reveals what we can—and cannot—uncover about everyday life in the past. Widespread cultural belief in olfaction's lack of history often goes hand in hand with an equally far-reaching belief about the demise of olfaction after the Enlightenment. The ephemeral remains of early modern perfume exist somewhere in between two conflicting historical approaches to the past; as such, they document a lost threshold of meaning, a zone of contact where air and bodies comingled. I examine this paradox, arguing that it turns on seventeenth-century redefinitions of bodily materiality. Olfaction, like other sensory ways of knowing, emphasizes the fungible relationship between material objects, the body, and embodiment. Its history thus represents both a limit and a threshold; it is a contact zone of subjective and social formations of knowledge. Contained within it is an archive of our collective desire for sapience and sagacity.

Cleopatra's Strange, Invisible Perfume

In order to create a counter-narrative to the history of perfume, one must first puncture the fantasy of perfume's exoticism and grapple instead with its ephemeral, invisible material influence. Thus, I begin with an early modern English fantasy about Egyptian perfumes and their effect on Roman bodies, Shakespeare's *Antony and Cleopatra*. Cleopatra's "strange invisible perfumes" are perhaps the best known in early modern English literature. As Enobarbus famously observes, Cleopatra is a master of theatrical effects, specifically multisensorial ones. In his oft-quoted recounting of the couple's initial meeting in Tarsus, Enobarbus describes how, even before Antony saw her, Cleopatra "pursed up" his heart "upon the river Cydnus" (2.2.192). Enobarbus describes her barge, with a golden deck, purple sails "so perfumèd that the winds were lovesick with them," and silver oars that beat to the tune of flutes (2.2.199–200). Fanned by dimpled boys, Cleopatra's visage "beggared all description," but Enobarbus notes that it was her perfume that first affected Antony: "From the barge / A strange invisible perfume hits the sense / Of the adjacent wharfs. The city cast / Her people out upon her . . . and Antony / Enthroned i'th' market-place, did sit alone, / Whistling to th' air" (2.2.217–19). Cleopatra's scents designate a wide sphere of environmental influence. Enobarbus's account contradicts Antony's belief in love at first sight expressed in the play's opening scenes; Antony fell in love not at first sight but at first smell (2.2.197–99).[77]

The setting underscores his claim: Tarsus, in Turkey, was the legendary home of King Caspar, one of the three Magi, who offered gifts of myrrh to the infant Christ. It was well known in the early modern period for its biblical associations with the Epiphany—and with perfumes (discussed more fully in chapter 1).[78] Summoned to Tarsus by the Roman imperium seeking to demand monies from wealthy Egypt, Cleopatra employs spices to exploit Roman assumptions about luxurious Egyptian exoticism, a strategy also used in mayoral displays of London's civic and economic power.[79] As Jonathan Gil Harris argues, Cleopatra's power on the Cydnus depends on the failure of narrative descriptions of visual splendor: "the curious lack of physical detail offered by Enobarbus about Cleopatra's displayed body suggests that her power subsists in her very invisibility, her publicly paraded absence."[80] Whereas Harris concludes that such invisibility underscores a scopophilic, narcissistic pleasure central to early modern theaters, perhaps such invisibility cites early modern theatrical practices—and pleasures—that relied on other sensory mechanisms, particularly "strange" and "invisible" scents.

Though it was enacted on an early modern stage, where "some squeaking Cleopatra boy" portrayed Shakespeare's queen, Enobarbus's description of Cleopatra's beguiling entrance in Tarsus underscores the theatrical power of her self-presentation (5.2.216–17). Juxtaposed with Octavia, who "shows a body, rather than a life" and is more "a statue than a breather," Cleopatra's sexual power (at least in Enobarbus's estimation) is expansive; hinged to the power of her perfumes, her influence extends beyond her immediate realm and works in subtle ways. Although Caesar's Roman "eyes" are on Antony, knowledge of his "affairs come to [Caesar] on the wind" (3.3.63–64).

To modern critics, Shakespeare's Egypt is an exoticized locale conflated with its queen.[81] Critics have described Shakespeare's Egypt as a "carnevalesque," "mercurial," "emasculating," "intoxicating," "timeless," and "overflowing" zone—Rome's antithesis—raising questions about how it may have been staged.[82] As theater historian Barbara Hodgdon notes, until recently, the spatial and geographic extremes presented in *Antony and Cleopatra* seemed unperformable.[83] Literary critic Jonathan Dollimore asks a similar question: "[Antony's] sexuality is structured by those very power relations he is prepared to sacrifice for his sexual freedom—Rome for Egypt . . . [b]ut how to convey this in production?"[84] Cleopatra's changeability reveals the limitations of vision in depicting her powerful sexual appeal on modern stages. The play's oblique olfactory references suggest that perfumes signified Cleopatra's power and desirability in ways visual codes could not capture.

On early modern stages, scent was often used to indicate cues of social difference, including gender and sexuality.[85] *Antony and Cleopatra*'s ending highlights this claim, first invoking Cleopatra's theatrical, perfumed excess then disciplining her for it. To control Cleopatra's—and Egypt's—atmospheric influence on vulnerable Roman bodies, the play must end with her sequestration. Barbour's interrogation of Shakespeare's "proto-orientalist" metaphors suggests rich interpretations of how early modern playgoers imagined the east, but to interpret Cleopatra's perfume solely as a marker of fantasies about eastern opulence obscures the material history of perfume in England and on English stages. It is, after all, hard to imagine that perfume *always* signified something "exotic" and foreign, especially since perfume had been produced, distributed, and represented in England since the thirteenth century. Perfumes were produced from a variety of ingredients and signified in multiple ways. Perfumes may have suggested Egyptian exoticism, but they also could represent a powerful (and perhaps English) queen in control of her environment.

Whereas commentators within the play repeatedly suggest that olfactory ploys reveal Cleopatra to be a cosmetic charlatan, witch, and whore, her influence is intoxicating, defining her kingdom as a zone of pleasure for Roman soldiers. Her tactics work; in the opening acts of the play, gender inversion, Bacchanalian excess, and sexual role-playing all seem to unman Roman martial prowess.[86] In contrast to the rigidity of Rome, Egypt's pleasures are polymorphous, enthralling Roman bodies with tantalizing desires.[87] Yet Cleopatra's scents are not named. Rather than list specific ingredients, Shakespeare describes Cleopatra's perfume in general terms as swift, strong, and intoxicating. Portrayed either through the locations in which they are encountered (a riverbank, the open seas, a tomb) or the bodily effects they inspire (drawing a city's population to the shore, bewitching Antony to abandon battle, or inspiring death) her perfumes define her powerful sphere of influence. By the end of the play, she mocks theatrical attempts to recreate her power, asserting that her scent is like death itself, "a sweet balm, soft as air" (5.2.311).

To name scent ingredients would disrupt the play's subtle lament for a queen in control of her body, representation, and environment.[88] Cleopatra's introduction in the play links her to air: Philo mocks Antony's fanning dotage of Cleopatra as she enters the stage, accompanied by eunuchs performing the same task (1.1.10). Later in this act, Enobarbus notes that her passions are equivalent to a geological force: her sighs and tears are "winds and waters" and "storms and tempests" (1.2.133–37). He marvels at her ability, when breathless, to make defect perfect and "pour breath forth" (2.2.238). Even in death, Cleopatra attempts to transform herself, shedding her feminine, humoral elements of earth and mud and willing herself into another realm—"I am fire and air." She partially succeeds. Charmian notes a changed atmosphere in the tomb: "Dissolve, thick cloud, and rain, that I may say / The gods themselves do weep" (5.2.280, 290–91). Cleopatra's "infinite variety" includes her deliberate manipulation of the air that surrounds her.

Though staged as Egyptian, Cleopatra's perfume (and Shakespeare's invocation of it) gesture to the literary and material significance of scents in early modern English culture.[89] What would Shakespeare's audience make of these olfactory details? Would they imagine her death as a moment of strength or weakness? Does the play celebrate or denigrate of the allure of female sexuality that hangs in its air? What is the relationship between perfume and early modern political networks of sex, gender, and imperialism explored in the play? These are literary questions; to answer them, however, requires attention to the material history of perfume so often ignored. How was perfume produced, applied, received, or represented in early modern England? What is the relationship between artistic representations of perfume and its production, distribution, and value in early modern English culture? And, more broadly, what is the role of olfaction in understanding the relationship between the practice of history and the production of cultural meaning? Or, to pose the question differently, how did early modern theatergoers imagine Cleopatra's perfume to smell? Enobarbus seemingly offers no clues. But for early modern English men and women, Cleopatra's strange, invisible perfumes signified not only poetically and metaphorically but also materially. Shakespeare's poetic description of Cleopatra's perfume is a starting point because it appears as a lacuna to a modern eye; it shows we have much to learn about what people thought of perfume's mysterious materiality, then and now.

CHAPTER ONE

Censing God

Frankincense, Censers, Churches

On June 17, 1440, eighty-year-old Richard Wyche was burned alive for heresy near the public execution scaffold on Tower Hill in London. Long associated with Lollardy, Wyche, an Oxford-educated priest, was tried and imprisoned for preaching Wycliffite doctrines as early as 1401; though he ultimately recanted, Wyche continued his preaching in the north during the early fifteenth century. By 1440, his associations with known heretics, including Sir John Oldcastle and William Thorpe, along with his popular evangelical sermons, were enough to charge him with obdurate heresy.[1] Wyche was tried, convicted, condemned, stripped of his priesthood, and finally, executed. Public reaction was swift and strong, provoking, as one chronicler of London reports, "moche trobil a-monge the pepylle."[2] Almost immediately, a cult of martyrdom and pilgrimage developed: his execution site was made into a miraculous shrine, marked by a large stone cross, where many left waxen images and offerings, burnt candles, and prayed. Reports described that a divine scent of incense issued forth from the ashes of his pyre.[3] Wyche's sweetly scented remains were soon sought as relics. One faithful follower transported them home in a pyx.[4]

Such devotional practices for a convicted heretic threatened civic order. Wyche's posthumous perfumes offered proof to his followers that he was no ordinary preacher; rather, he was a true martyr, challenging the sanctimonious authority of his executioners. Authorities quickly countered Wyche's air of sanctity with an olfactory hazard of their own, establishing a dunghill on the execution site to dissuade the faithful from transporting any more of his remains.[5] On June 23, Henry VI and his bishops ordered the lord mayor to act against breaches of the peace in the city and its suburbs, summoning all absent aldermen to help.[6] By July 14, authorities turned their attention to those who encouraged the cult, arresting several Londoners, including Thomas Virley, vicar of All Hallows Church, Barking, for perfuming Wyche's remains. Virley was accused of "receiv[ing] the offerying of the simple peple," and of taking Wyche's "ashes and medlid thaym with powder of spices" so that "the simple peple was decyved, wenying that the swete flavour hadde commed of the asshis of the ded heretic."[7] Using liturgical scents or perhaps spices available in local London markets, Virley had manufactured the smell of divinity.[8]

Wyche's fragrant and foul remains demonstrate that, in fifteenth-century London, there was more than one way to smell divine. Though incense was routinely utilized in late medieval liturgies to signify the presence of divinity, it could also be smelled in other spaces as more and more spices were available in England. What happened when heavenly scents (or, for that matter, sounds, sights, tastes, or touches) were encountered on Tower Hill, rather than in a church or cathedral? How did one know when one was in the presence of God and one was in the presence of something more earthly? Could one smell which was Wyche, a martyr or a heretic?[9] Using a standard Catholic trope of devotion—the odor of sanctity—Virley's olfactory stage effects represented both the allure and the limits of such sensory cues. Virley's ability to manufacture the odor of sanctity through earthly means, along with the authorities' efforts to counter it, demonstrates the important role olfaction played both in late medieval Catholic devotional practices and in early Protestant protest against them.

If Virley's perfuming represents the power of Catholic belief in the odor of sanctity, then William Caxton's report represents the Protestant counterpoint. Caxton had witnessed Wyche's execution. Writing about it fifty years later, he emphasized not Virley's perfumes but rather the authorities' olfactory response to them. Emphasizing the dunghill as part of a series of degradations inflicted on Wyche, Caxton chronicles that Wyche died a good Christian: "This yere Syr Rychard Wiche, vycary of hermettesworth was degrated of his prysthode, at powlys, and brente at toure hylle as for an heretik on saynt Botulphus day, how wel at his deth, he deyed a good crysten man, wherefor after his deth moch peple cam to the place where he hadde ben brente, and offred and made a heepe of stones, and set up a crosse of tree, and helde hym for a saynt till the mayer and shreves, by commaundement of the kynge and bisshops destroyed it, and made there a donghyll."[10] Caxton's narrative, oft-quoted and cited in later sixteenth-century chronicles, highlights visual and olfactory symbolism: the heap of stones and the cross symbolize Wyche's saintliness until they are destroyed and replaced with a dunghill. Such sensory ways of knowing contributed to folk recognition of Wyche's martyrdom at the "commaundement of the kynge and bisshops."

Though the olfactory politics at work in Wyche's tale may seem obvious (sweet scents represent his divinity, and scents like sulfurous dung hint at something foul), Caxton's reconfiguration of them demonstrates that such links were not intrinsic; rather, they were forged through almost a millennium of Catholic religious devotional practice. Olfactory descriptions of late medieval Catholic martyrdom, for example, like the icons of early Christian saints, could have emphasized the tortures inflicted on martyrs' bodies. Indeed, up until the fourth century, such descriptions usually did. Set against Roman use of perfume, descriptions of the deaths of early Christian martyrs stressed their willingness to suffer earthly horrors (including olfactory degradation) for their faith. Though many of these accounts mention perfumes, especially those like myrrh associated with embalming, such scents played up martyrs' human sacrifice rather than their divinity.

Within early medieval Catholic devotional practice, however, one scent transcended the earthly realm and materialized the presence of the divine—frankincense. Produced from spices originating in the Holy Land and supplemented with native botanicals, religious perfumes were the product of a myriad of geographic, political, and economic networks. The medieval Catholic liturgy used perfumed incense to transform parishioner's prayers into a sweet substance that could transcend the earthly realm and rise up to God. Deriving from ancient Judaic and Arabic traditions, the religious use of incense first appeared as part of the Catholic liturgy in the fourth century. As church architecture evolved and expanded, incense remained a key signifier of religious transformation. Perfumes were thus a key component of early Christian religious experience, yet their meanings at that time were fundamentally different from late medieval ones.[11]

Medieval Catholic prayer involved a number of sensory experiences and attempted to create proximity to divinity. People prayed in front of images of saints, burned incense and candles, touched relics, and chanted. For most medieval men and women, such divine proximity likely occurred only twice a year during the celebration of Good Friday and Easter Sunday masses, the apex of the medieval church calendar. The Passion prepared the piety for Easter and a celebration of literal proximity to Christ's body through communion. Incense was part of both services; it also was used in liturgical dramas that marked the transformations between the two ceremonies.[12] By the twelfth century, incense was an established part of the liturgy, symbolically linked to holiness in general and the sacraments in particular, especially communion. The visual splendor of the host was amplified by ritual censing during processionals. Incense was also used as part of services for consecrations of churches, cemeteries, and anchorites; translations of saintly relics; ordinations of priests; holy matrimonies; and the administration of last rites.[13] By the fourteenth century, ritual censing defined such practices, filling church spaces with smells that marked divinity.[14] As Virley's perfuming of Wyche's remains demonstrates, by the fifteenth century, incense was a powerful way to represent the presence of the divine through earthly means.

However, Virley's olfactory trick also shows that perfumes were no longer a rarified scent found only in churches. The sweet scent of incense competed with the scented allure of ambergris, spikenard, musk, civet, pepper, clary sage, grains of paradise, ginger, and cinnamon increasingly found in England's markets; similarly, early reformers routinely commented on what they believe to be the "lewd" use of liturgical scents by the laity. The ritual use of incense in the medieval liturgy transformed prayer, objects, and spaces into holy realms. But Wyche's saintly martyrdom makes clear that when new kinds of aromatics entered England's markets in the late fifteenth century, the olfactory markers of divinity began to signify other kinds of transformations, namely England's participation in a global spice trade. The exotic, the profane, and the holy all were invoked by scented incense and balm. As such, late-fifteenth- and early-sixteenth-century representations of religious olfactory experience often squarely address such confusion, asking when was a scent celestial and when was it profane?

England's late medieval Catholicism and its early reform movements offer pedagogical approaches to such questions. The material history of incense, as a product of global exchange and tool of religious epistemology, is embedded within fraught debates over what defined a "good" Christian in fifteenth- and early-sixteenth-century England. Religious dramas, which sought to educate and entertain the laity, offer one way to approach this history through both global networks of trade and personal, sensory experience.

Sensing God: Incense and Metaphors of Prayer

The story of Wyche reveals how late-medieval men and women conceptualized their faith through direct sensory engagement with the material world. If Wyche's remains smelled sweet, he was sanctified; if they smelled like a dunghill, he was not. The power of such a test resided in the sanctified object's ability to transcend its environment. Late medieval descriptions of the odor of sanctity often relied on olfactory markers encountered in environments other than church.[15] After Thomas Becket's murder, for example, Christ's Church was overwhelmed with the smell of scented ointment; one witness claimed it was as if a secular perfume box had suddenly been opened. The scent lingered, despite the fact that monks, hastening the burial process, did not have time to properly clean, incense, and embalm his body.[16] St. Wulfstan's tomb, opened in the twelfth century, also poured forth scent.[17] In these descriptions, sanctity smells like earthly perfumes rather than the other way around. John Mirk's *Festial,* a fifteenth-century collection of homilies, likened the scent of the Virgin Mary to that of a spicer's shop, "for as spycers schoppe smellþe swete of diverse spices, soo scho for þe presens of þe Holy Gost þat was yn hur and þe abundance of vertues þat scho smellyth swettyr þen any wordly spycery"[18] [for as spicers' shops smell sweetly of diverse spices, so too did she, for the presence of the Holy Ghost that was in her and the abundance of virtues did smell sweeter than any worldly spicery]. Such descriptions indicate how secular perfumes metaphorically produced associations with sanctity through the effect of their scents on the surrounding environment.

Not surprisingly, early reformers strove to demonstrate—and divorce—the material connections between sanctity and such earthly scents. Lollards attacked such multisensorial effects as too sensuous, in particular aromatic offerings to "blynde rodys and to deue ymages of tre and of ston"[19] [blind roods and deaf images of wood and stone]. William Thorpe, a Shrewsbury priest arrested in the early fifteenth century for preaching Lollard doctrines, argued that no one should make vows to saints "nor seek them nor kneel to them nor kiss them nor incense them."[20] Other fifteenth-century reformers sought to limit the use of incense to liturgical ritual: as a tool of divine prayer, it was too powerful for man's hands and should only be used by a priest at an altar.

Dives and Pauper, an early-fifteenth-century treatise, warns against incense's use by "lewyd people." In this text, Dives, an aristocratic male, enters into a dialogue with Pauper, a poor, itinerant preacher, about the Ten Commandments. Dives describes his outrage over the widespread practice of censing as a component of prayer, warning that such behavior reeks of idolatry: "[T]huryfication and encensyng was be held tyme an heye dyuyn wurshepe, and manye seyntes weryn put to the deth for they woldyn nought encensyn ymages, sotckys and stonys. But now clerks censyn the images, and other preistys and clerkys and the lewyd peple also, and so, as me thynkethy, they don ydolatrie."[21]

Dives identifies incense as a cause of early Christian martyrdom. Pauper, however, answers that censers are powerful tools if used correctly: they translate prayers from human hearts into sweet-smelling incense that can transcend earthly realm and reach heaven. The class hierarchies structuring their debate demonstrate how routine access to material perfumes shaped its religious meanings. Dives is horrified by perfume's ability to transcend liturgical use, so much so that "lewyd people" use it in unspecified ways, whereas Pauper emphasizes its important pedagogical value.

Such a debate demonstrates that, as the perfume market expanded in the late fifteenth and early sixteenth century, incense, myrrh, and other aromatics signified in more earthly ways. For early reformers, censing became an allegorical abstraction rather than a metaphorical reference to real-world smells. Troubled by the association of prayer with such worldly goods, Erasmus argued that prayers were purer than incense, myrrh, or galbanum (another biblical perfume): "Pure prayers also and geuynge of thankes do make to god a perfume more pleasante / than any incense myrhe and galbanus."[22] [Pure prayers and a giving of thanks do make to God a perfume more pleasant than any incense, myrhe and galbanus.] In other words, prayer is a metaphorical perfume one can produce to please God.

Increasingly, the purity of prayer depends upon deodorization, removing both its past history of sacrifice, signified through Jewish materiality, and its present "lewdness," signified through its associations with wayward priests or with Islam. For example, Thomas Becon, in his many treatises, links incense with a Jewish materiality, signifying not the power of prayer but rather violent, corporeal sacrifices of the Old Testament.[23] In Chaucer's "The Miller's Tale," incense has more sexual connotations. Absolon, the parish clerk, first notices Alison while censing in church: "This Absolon, that joly was and gay, / Gooth with a cencer on the haliday, Cencing the wives of the parish faste, / And many a lovely look on hem he caste." Absolon's priestly, and promiscuous, "looks" are linked to his ritual censing of parish women on the "holiday."[24] And Martin Luther vociferously attacked incense's association with luxury and sexual sin, denouncing all "catamites and Ganymedes" who destroy the Church, including "the effeminate Arab" who "sells balls of scented incense."[25]

Despite their overt engagement with broader structures of trade in the circum-Mediterranean world (specifically with Judeo-Islamic trade networks in the Levant), attempts by such representations to link the scent of perfume with sin have been understood as part of a broader Protestant attack on the sensuous materialism of Catholicism. Eamon Duffy's influential history of England's early reforms famously argued that the stripping of Catholic altars during early Protestant reforms sought to counteract sensory ways of knowing God. Such divestments were grouped under broader attacks against "idolatry," defined as any practice that conflated the tools used to venerate an unseen or invisible deity with worship of materiality itself. Incense, one might imagine, would factor largely in such divestments. It was, after all, a key component of liturgical ritual, amplifying visual and aural tropes during the Catholic mass. But the earliest Protestant reforms in England are curiously silent on the practice of censing. It was not until the third injunction of 1547 that incense was mentioned at all, requiring all images that had been censed be taken down.[26] An official change may have been made after the 1549 Acts of Uniformity, though they make no explicit mention of incense.

In his study of the use of smell on England's seventeenth-century Protestant stages, Jonathan Gil Harris argues that by the seventeenth century, religious perfume, specifically incense, was emptied of all symbolic meaning: "after the Reformation, all that remained for the nose in religious representation was the foulness of the diabolical."[27] Harris has persuasively demonstrated how such olfactory foulness manifested itself on the stage, focusing on the ability of "squibs" to "shatter the olfactory coordinates of Protestant time."[28] The abolition of censing in England's churches, however, was slow and uneven. St. Michael's in Gloucester, for example, purchased coals for incense in 1547 but not in 1548.[29] In 1563, at a church in Tewkesbury, three men protested the use of Catholic ritual, including a number of "not defaced relics," such as candlesticks, censers, incense, clappers, processionals, and two mass books.[30] The use of incense continued during the early seventeenth century. Records indicate that in 1603, two pounds were burnt in the Church of Augustine, Farringdon, and that in 1626, £2 worth of incense was burnt in Great Wigston, Leicestershire.[31] Prynne's satirical *Canterburies Doome* mocks both James I's and Charles I's use of incense.[32] As late as 1683, frankincense was used during the procession of a Bishop at St. Nicholas, in Durham.[33]

Whereas Continental reformers attacked the use of incense as a luxurious, sexual sin, English reformers used incense as a metaphorical abstraction in their sermons.[34] In his sermon on Wycliffe, for example, John Bale praised the preacher's steadfast commitment to the "truth" by comparing him to incense burning in a fire.[35] Likewise, in his vision of Judgment Day, he creates a divine link between Abraham, John the Baptist, and all faithful followers on earth through Christ's divine censing.[36] The angel's censer collects the incense of the faithful, which, reignited by the fire of God's true altar, transforms into a Pentecostal fire for all of God's apostles. For Bale, incense is a metaphor of religious faith, mapping divinity through an osmology of purely divine scent. The actual practice of "sensyng," or burning of incense, however, was just another Catholic ritual involving "heythnythe wares" offered to the panoply of saints— "whoremongers, bawdes, brybers, ydolaters, hypocrytes, trayters and most fylthye Gomorreanes, as godly men and women . . . haue canonysed."[37]

In *A Priest to the Temple, or the Country Parson,* George Herbert, writing almost a century later, also invokes incense as metaphor for Protestant purity. Describing the death of Mr. Hullier, Vicar of Babram, who was "burnt to death in Cambridge," Herbert emphasizes that Hullier held a "Common-Prayer Book in his hand, in stead of a Censor" and used his "prayers as incense," offering himself "as a whole Burnt Sacrifice to God."[38] Protestant sacrifice and prayer emerge as true censing, displacing the abject materiality of the censer. Elsewhere in the text, Herbert advises country parsons to "perfume" churches "with incense" during "great festivals."[39] Such practices suggest incense's symbolic meanings were hardly emptied in the seventeenth century and remained complex.

In more than two thousand years of use in religious rituals, incense has acquired many symbolic meanings. For ancient Romans, its scent was just one of the many aromatics that saturated daily life. Both foul and fragrant scents were involved in Roman religious sacrifices.[40] In ancient Jewish and Egyptian religious practices, the burning of incense signaled a profound material transformation rather than an act of destruction.[41] Transforming earthly material into an ethereal, yet perceptible, fragrant smoke, the burning of incense symbolized a powerful transformation, key to rituals of sacrifice, prayer, and purification.[42] Early Christian practices at first sought to eschew such rituals, emphasizing instead odors of bodily sacrifice and inverting ancient associations between good smells and moral goodness. Yet as Catholic theology increasingly emphasized a direct, experiential link to God through sensory ways of knowing, holy oils and smokes like incense forged such pedagogical links.

The material history of incense, however, complicates the history of its symbolic power. Ancient Egyptians and Jews, classical Romans, and medieval Christians all may have valued "incense," but the term itself includes a world of distinctive scents from vastly different locations. Frankincense is a natural oleo-gum-resin harvested from a number of trees native to the *Boswellia* genus. Ancient frankincense was believed to derive from *Boswellia papyrifera,* native to east Africa, whereas medieval frankincense was composed of resins from *Boswellia sacra,* first discovered in Arabia (now modern-day Yemen, Oman, and Socotra). Many believe that its scent was far superior to its classical predecessor, though it may have been adulterated with cheaper ingredients like mastic, copal, or even English pine and sold under different names.[43] *Boswellia sacra* is now incredibly rare due to the gradual desiccation of the Arabian peninsula, so most modern frankincense is now harvested from *Boswellia carterii,* a small shrub native to Somalia and Ethiopia, whose scent is more similar to ancient frankincense than to medieval versions.

Frankincense is obtained by making small incisions into the bark of the tree, which produces a milky, pale-yellow gum that is scraped from the bark. When exposed to air for three months or longer, this gum hardens into dark, bumpy "tears," or pea-sized globes.[44] Frankincense was a popular luxury trade item in the ancient and early medieval world. Its alternate name, olibanum, from the Arabic "al lubán," or milk, marks the Levant as its source (and even provides Lebanon with its name).[45] Its Middle English name "frankincense," from the old French *"frank encens"* or "quality incense," emphasized the resin's scent rather than its appearance. The Latin term, *incensum,* or "that which is set on fire," or *thus,* from the Greek θύóος [sacrifice], described its use. (The liturgical term *thurible* derives from the latter.)

Medieval frankincense was most often paired with myrrh, a pungent, bitter scent (from the Hebrew *mōr* for "bitter") noted for its embalming applications. Mentioned a hundred times in the Vulgate bible and eighty-one times in descriptions of religious worship, medieval incense represented both Christ's human suffering and his divine transformation. In late medieval Christian beliefs, frankincense and myrrh, as material substances whose scents symbolized the dramatic power of both the Epiphany and the Resurrection, framed Christ's birth and death. Frankincense and myrrh were two gifts of the Magi to the infant king. Frankincense signaled Christ's divinity, whereas myrrh marked his corporality, foreshadowing the divine sacrifice of the Passion. In liturgical and civic dramas of the Passion, myrrh signals sacrifice. The three myrophores, or myrrh bearers, are the first to witness Christ's divine resurrection; their embalming scents dramatically emphasize his triumph over human death. In such representations, incense becomes a key marker of transformation and conversion to Christianity.

For anxious London church authorities, Richard Wyche's falsified sanctity may have underscored the paradox of divine epiphanies and their celebration within Catholic rituals: most were constructed from a very earthly materiality. As medieval miracle and saints' plays show, late medieval English men and women conceived of divine manifestations within human realms in striking sensory ways.[46] These texts illustrate the powerful connection between everyday olfactory experience and divine transformations. But they also complicate such connections, using incense and olfaction as a theatrical effect in staging divinity. Two plays, both featuring important religious figures defined by their perfumes, reveal the fault lines of censing and sensing, the Wakefield *Offering of the Magi* and the Digby *Mary Magdalene.*

Staging Incense: *The Wakefield Offering of the Magi*

The feast of the Epiphany was an opportunity to frame how divine signs were to be understood and interpreted as tools for worship, particularly as more and more scent ingredients from the Levant (integral to such church rituals) were available in England. One religious play, the Wakefield *Offering of the Magi*, demonstrates how one's faith shapes sensory perceptions of the smell of divinity. Given the popularity of the Magi and the important role of their offerings in the nativity story, it is almost impossible to imagine that the Wakefield play (or indeed any of the mystery guild productions representing the Magi) was staged without using incense as a stage property.[47]

The Magi, as individual characters, were strikingly heterogeneous, yet as a group they symbolized the potential for universal Christian conversion.[48] Though they were incredibly popular within late medieval devotional rituals, the Magi were not saints.[49] By the late eleventh century, they collectively functioned as pedagogical models of divine adoration rather than subjects worthy of such devotion.[50] Such adoration manifested itself through their material offerings of gold, myrrh, and incense to the infant Christ. The Magi (and their gifts) also became associated with the Levant. Matthew's gospel describes only the visit of *"magi ab oriente,"* an unnamed number of wise men from the orient.[51] By the late Middle Ages, however, names and provenance were ascribed for three Magi: Caspar, Melchior, and Balthasar, it was said, all came from near the border of Persia and Chaldea. The translation of their relics from Constantinople to Milan and then to Cologne dramatically heightened devotion to their tale in northern and southern Europe.[52] By the first decades of the twelfth century, their pilgrimage from the East to Bethlehem had come to serve as a model for those in the West, inspiring the kings of the first crusade to liberate the holy land from Herod-like figures.[53]

Whereas the Magi, as characters, represented the universality of Christian conversion, their gifts symbolized Christ's unique status as the human son of god. Frankincense represented his divinity; gold, his status as the King of all kings; and myrrh, his human mortality. Bede's history relates that the gift of gold relieved Mary's poverty, the incense sweetened the smell of the barn, and the myrrh purged Christ of worms, strengthening his infant body.[54] Voragine, in the *Golden Legend*, elaborates on these three "attributes," linking them to a host of significations: while gold was offered for tribute, incense represented Christ's ultimate sacrifice, and myrrh, his burial, "so these three things corresponded to

Christ's royal power, divine majesty, and human mortality." Gold, Voragine notes, symbolized both "love" and, "being the most precious of metals," the "precious divinity of Christ," while incense symbolized "prayer" and Christ's "prayerful soul." Myrrh symbolized "the mortification of the flesh," which "prefigured his uncorrupted flesh."[55] As their centrality within celebrations of the feast of the Epiphany and Twelfth Night developed, so, too, did the role of Herod in the tale, both as a comic symbol of misrule and as an obstacle to faith. His theatrical rants became an integral part of the season's comic emphasis on misrule, his worship of Muhammad complicating the relationship between the Magi's journey from the east and their gifts to the infant messiah.[56] The Wakefield *Offering of the Magi* highlights this link in its staging of olfactory offerings.

The play opens with Herod's command of silence, a theatrical ploy that blurs boundaries between the social world of the play and that of the stage. Declaring himself a lord "of all this warld," Herod anachronistically aligns his power with Mohammad's, declaring that "to Mahowne and me all shall bow" (line 15).[57] Belligerently declaring war on all "faturs . . . that will not trow on Sant Mahowne, / Oure god so swete," Herod's opening speech blurs temporal distinction between biblical narrative and contemporaneous politics and culture. Further, he implicates the audience in worship of this "sweet Sant," inversely linking Islam to Christian olfactory practices of religious worship.

Though the play ostensibly stages the Epiphany as a visual triumph over such heretical inversions, it also shows incense to be the proper scent associated with "sweet" divinity (line 185). When the kings enter, they meditate on the "mervel" of the star (line 201), demonstrating how a divine, visual spectacle can interrupt the earthly plane and, most likely, the dramatic stage as well. The kings (and perhaps the audience) struggle to interpret such a sign: is it an astrological anomaly (Jaspar), a novelty and marvel (Melchor), or part of prophecy (Balthesar)? Then each grows convinced that it is a prophecy of divinity, a portent of prince's birth "that shall overcome in hy Kasar and King" (line 220). Their individual pilgrimages converge into a singular act of worship: each offers a token fitting for such a prince. Jaspar offers gold, as a "tikyn that he king shal be / Of alkyn thing." Malchor brings "rekyls, . . . in token that he God veray / withoutten ende." Balthesar offers "myrr," "in token that he shal be ded" (lines 229–46).

Malchor's offering of "rekyls," or frankincense, emphasizes his gift's smoky exhalations. *Rekyls,* from the French *récan,* "to reek," was another term for "ansens," or incense, from the Latin *incesum,* or "that which is set on fire." Malchor's rekyls symbolizes the unity of the trinity, a divinity "veray withouten ende" through the transformative power of burning resins in order to produce sweet smoke, a reference that directly challenges Herod's worship of Mahowne's "swete" sanctity earlier in the play (lines 239–40). In case an audience member might have missed such a connection, the play offers another olfactory counterpoint almost immediately. Spying on the kings, Herod's servant Nuntius abandons his original task—to survey the audience and all believers in the land—in order to report to Herod the arrival of the kings. Herod greets Nuntius with an olfactory slur: "Where has though bene so long fro me, / Vile stinkand lad?" (lines 261–62). Nuntius's stench undoubtedly functions as comic relief. Yet it also reveals that Herod's ability to discern divine fragrance is woefully inept: Mawhone's sanctity is "swete," while Nuntius and his report of the kings and their prophecy offers a vile, stinking stench.

Rather than pantomime the presentation of the gifts, the play dramatizes how one should worship Christ, linking Malchor's incense with pedagogies of sensation. As each king offers his gift, he recites a prayer that begins with a direct address to Christ: "Hail be thou, maker of all-kyn thing," "Haill, overcomer of king and knight," and "hail, king in kith, cowrand on kne! / Haill, oone-fold God in persons thre!" (lines 541, 557, 553). When linked with the visual and olfactory properties of such gifts, the characters' apostrophes mimic both liturgical prayer and liturgical offerings, infusing the scent of divinity with medieval street theater.

The play's stress on olfactory sensation is heightened when read against other plays in the Towneley cycle that dramatize the nativity, namely the *Second Shepherds' Play*, which would have immediately preceded it if performed as a complete cycle. Whereas the *Offering of the Magi* emphasizes the olfactory as a key part of divine adoration, the *Second Shepherds' Play* reveals the limits of such olfactory stage effects. Though Herod claimed his god was "swete," the Magi instead offered fragrances that, like Christ's divinity, transformed and transcended earthly materiality. In the *Second Shepherds' Play*, the olfactory equally reveals the limits of earthly machinations. Gyll claims that labor has altered her ability to bear the scent of the shepherds and the nearby moors: "Go to anothere stede! I may not well qweasse; / Ich fote that ye trede goys thorow my nese / So hee" (lines 487–89). Yet Daw remarks that her child smells as "low" as livestock: "Whik cattell bot this, tame nor wylde, / None as have I blys, as lowed as he smylde" (line 549), which, of course, it is.[58] The central joke of Mak and Gyll's ruse, while seemingly blasphemous, emphasizes that these two "barnes"—the stolen sheep and the infant Christ—are not the same: one is the lamb of god, the other, merely a lamb. Olfaction, like an ill weft, or evil, will out: one smells like god, the other, a lamb.[59]

Smelling Sin: Incense and Mary Magdalene's Perfumed Conversions

Theatrical performances like the *Second Shepherds' Play* suggest the paradoxes latent in late medieval English approaches to perfume. Though it offers a clear logic of olfaction mapped onto sin—god smells divine whereas animals do not—the fact that bad smells could function as stage properties, along with Malchor's staged offerings of "rekyls," complicated the meanings of divine smell. Incense was thus key to understanding the politics of censing and sensing in late medieval England. Was incense a perfume rather than a marker of divinity? If so, could it signify sin in similarly sensuous ways?

Perhaps no other figure better represented the inherent paradoxes of perfume's saintly and sinful connotations than the late medieval Magdalene. Medieval understandings of Mary Magdalene and her saintly transformation were linked with experiences of perfume; both liturgical and secular drama routinely staged her sinful past, her proximate and penitential relationship to the body of Christ, and her saintly conversion through perfumed stage properties, including incense. Whereas incensed processionals through the space of a church (which, by the thirteenth century, included dramatic performances of Mary Magdalene's visit to the holy sepulchre as a myrophore) prepared the laity for Christ's dramatic triumph over death, English medieval saints' plays like the Digby *Mary Magdalene* used olfactory cues to demarcate the saint's sinful sexual past and demonstrate her triumph over it.

The Magdalene is the only female saint who is not a wife, virgin, martyr, or mother.[60] Her ability to represent the power of temptation as well as redemption may explain her widespread appeal in the late medieval period. Both aspects are conflated in her personal attribute of perfumed balm. In medieval depictions, the Magdalene was identified through several attributes: a red cloak; long, loose hair; tears; a book; and an ointment bottle. These visual clues help identify her as a reformed prostitute—the red garment and unruly hair indicating her past promiscuousness, the book and tears symbolizing her repentance.[61] The ointment bottle containing scented oil or balm works as a hinge between these two aspects of her character and thus is critical to understanding (and staging) Mary Magdalene's descent into cupidity and her penitent submission at the feet of Christ.

In the Digby play her tale of sexual and religious transformation is staged through contemporaneous politics and trade. Little is known about the play's author, its date, or its production history, though scholars estimate it was first performed between 1480 and 1530 and that it can be assigned to East Anglia.[62] With more than two thousand lines and possibly as many as fifty-two scene changes, this secular play is, in Theresa Coletti's summation, "awesomely eclectic" and "contains nearly every theatrical device known to the late medieval playwright."[63] Clifford Davidson argues that its sprawling structure and seemingly extravagant use of props suggest that it was probably a town production, most likely sponsored by a guild in honor of the Magdalene's feast day, July 22.[64] Scene descriptions indicate numerous costume changes; elaborate machinery, including a large ship that could travel back and forth across the stage and a device that could enable devils to exit the Magdalene's mouth on cue; and props to simulate the banquet and tavern scenes, including a plentiful supply of "awromatics."

Based on both the Golden Legend and the Gospels, the play depicts the biography of the Magdalene and balances the demands of church dogma against the guild economics that most likely financed its production.[65] As one scholar notes, the play "is as composite in structure" as the "Magdalene figure is in nature."[66] Organizing the saint's life into a dramatic and coherent triptych, it uses perfumed incense as a bridge between the first half's enactment of sin and repentance and the second half's staging of the miraculous conversion of the king of Marcyll. In several important "perfuming" scenes, it stages both the sinful nature of a luxurious touch and the venerating touch of salvation through the manipulation of Mary Magdalene's saintly emblem.

The play thus participates in the late medieval fascination—and unease—with this popular saint's complicated role as both sexual sinner and penitent role model.[67] As Coletti argues, the Digby play was part of a broader cultural interest in maintaining feminine virtue through social order, which was evidenced in church dogma, vitae, and popular conduct books. These documents sought to "explain, though not excuse, the Magdalene's associations with sexual vice," exploring "how the figure who was understood as Christ's intimate follower and beloved could have been guilty of sexual sin."[68] The Digby play explores this question through staging proper and improper engagement with worldly goods. The repeated utterances of "Oh, Mahommed!" in the opening scenes situate the Magdalene's descent from virtue anachronistically within a late medieval political geography of trade and empire. Within such a temporal displacement, the play's many calls for "wine and spices" mimic structures of trade symbolism, serving as tools of celebration within the play's narrative as well as stage properties culled from medieval markets. Likewise, the play's many meditations on the smell of spice ingredients educate its audience about navigating the pungent and sweet scents of the medieval church and market.

It is, after all, the diabolical King of Flesh who convinces Mary to leave her father's home and enter the social realm of the tavern where she is debauched. His is a floral, spiced dominion:

> I, Kyng of Flesch, florychyd in my flowers, Of deyntys delycyows I have grett domynacyon! So ryal a kyng was neuyr borne in bowrys, Nor hath more delyth, ne more delectacyon! For I haue comfortatywys to my comfortacyon: Dya galonga, ambra, and also margaretton— Alle þis is at my lyst, aȝens alle vexacyon! All wykkyt thyngys I woll sett asyde. Clary, pepur long, wyth granorum paradysy, Zenzybyr and synamom at euery tyde— Lo, alle swych deyntyys delycyas vse I![69]
>
> [I, King of Flesh, flourishing with flowers with dominion over such dainty delicacies! So royal a king was never born in any bower, nor has more delight, or more delectation! For I have comforting medicines that strengthen: galingale, ambergris, and also marjoram— all this is my desire against vexation! All wicked things I will set aside. Clary, pepper, grains of paradise, ginger, and cinnamon at every tide—lo, I use and delight in all such dainties!]

He is accompanied by a sparkling "lady lechery," "glittering" with "amorousness," referring, undoubtedly, to the greasy, hot, savory nature of the King of Flesh's scents.[70] Filled with desire for the king, Lady Lechery has shiny skin that invokes the materiality of the King of Flesh's "comforting medicines," specifically greasy ambergris. His scents—clary, pepper, grains of paradise, ginger, and cinnamon—work as medicinal ingredients against vexation through emitting (and enacting) sensuality and odiferous pleasure. These ingredients, not unlike those used to "spice" Wyche's remains, were commonly imported and sold in the late fifteenth century by the Company of Grocers. Coming from oceanic, northern European, Egyptian, African, and English sources, they demonstrated that sensual delight could be found in global perfumes.[71] They undoubtedly resonated as both foreign and familiar at the same time.

Lady Lechery, also referred to in the manuscript as "Luxuria" and "Luxurysa," is thus a character who represents both worldly and sexual temptation. *Luxuria* was a generic term for luxury goods as well as an ecclesiastical legal term of art meant to denote both gluttony and a hierarchy of other sexual sins, beginning with heterosexual sex outside of marriage and progressing to incest, sex with priests, and crimes against nature.[72] Luxuria was commonly depicted as two lovers embracing or as a rote depiction of debased feminine sexuality. Although the play enacts the Magdalene's descent into vice as a heterosexual liaison, Lady Lechery enacts the sinful temptation latent in both her worldly and sexual allure. She introduces the Magdalene to the tavern and its many sexual temptations. Described by the King of Flesh as a "flowyr fayres of femyntye," Lady Lechery's beauty mirrors the Magdalene's (described earlier in the play as "ful fayur and ful of femynyte"). Escorted by a bad angel, Lady Lechery breaches the Magdalene's castle. The scene is suggestively erotic, linking female beauty with vanity and vice.[73] The Magdalene is seduced by Lady Lechery's amiable tongue: "Your debonarius obedyauns ravyssyth me to trankquelyte! Now, syth ye desyre in eche degree, To receyve youw I have grett delectacyon! Ye be hartely welcum onto me—Your tong is so amyabyll, devyded wyth reson."[74] [Your debonair obedience ravishes me to tranquility! Now, since you desire in each degree, I have great delight to receive you! You are heartily welcome—Your tongue is so amiable, measured with reason.] Lady Lechery's greasy, sparkling amorousness metaphorically and materially signifies the power of scent to penetrate geographic and bodily boundaries.

Despite the Magdalene's sexual contentment in the floral bower with a young gallant (Curiousity), a "Ghost of Goodness" quickly convinces her to search for a salve for her soul; under his counsel, the Magdalene pursues the Prophet, seeking to trade her erotic, sweet balms for "an oil of mercy," dramatically enacting the medieval notion of *contrapasso,* a structure of penitence demanding that any absolution match the nature of the sin.[75] The play dramatizes her anointing of Christ's feet as part of his parable of the two debtors. When read on the page, it is not an erotic scene; Mary's wretchedness provides a useful illustration for Christ's tale of debt and forgiveness. Some critical editions of the play, however, interpolate Simon's shock at Christ's reception of the Magdalene into the text of the play, inferring that medieval audiences would recognize that the penitent Magdalene had been contaminated by Lady Lechery and any subsequent physical touch from her would spread the sin of luxuria. If staged with scents, this small dalliance within the broader pedagogical narrative could dramatically enact contrapasso, complicating the play's "smellscape" of sin and saintliness. The sweet balm conjures earlier scenes of the sparkling ambergreased amorousness of Lady Lechery, the fragrant bliss of the bower, and the Magdalene's search for an aromatic salve that could anoint and save her soul. As stage properties, such smells would further complicate the saintly and sinful meanings of perfume, particularly given the Magdalene's relationship to Christ.

Lest there be any doubt about the Magdalene's handling of Christ's body, the play indicates that, once penitence is achieved, the Magdalene must abandon her sweet balms for churchly incense. When she attempts, once again, to approach the resurrected Christ with her sweet balms, Christ issues forth the vernacular command "touch me nott" and offers her a didactic lesson in olfactory abstraction:

he is now like a gardener, tending a "gardyn" of humanity that "is watered with tears" and "springs virtue" that smells "full sweet."[76]

This abstraction is spun out even further. Christ provides yet another description of scent, this time on the proper scent of a woman: the Virgin. Almost halfway through the drama, Christ addresses the audience from the heaven stage and meditates on the blessedness of the Virgin. Lamenting that his mother's virtue cannot be described by language, he describes her instead through an epistemology of scent. In a play that stages Mary Magdalene's path toward sainthood, Christ's meditation on the Virgin may seem surprising, and in fact, it is the only moment in the play in which the Virgin is mentioned. She is incense, cinnabar, musk, and gillyflower.[77] "My blyssyd mother, of demvre femynyte . . . She is the precyus pyn, full of ensens, The precyus synamvyr, the body thorow to seche. She is the mvske aȝens the hertys of vyolens, The jentyll jelopher aȝen cardyakyllys wrech. The goodnesse of my mothere no tong can expresse, Nere no clerke of hyre, hyre joyys can wryth."[78] [My blessed mother, of demure femininity . . . is a precious pyx, full of incense, the precious cinnabar her body through to seek She is the musk against the hearts of violence the gentle gillyflower against such cardiac pain. No tongue can express her goodness. No clerk for hire can write her joy.]

The description of the Virgin's scent is a material olfactory experience—none can "write" her joy. It also marks a crucial transition in the tale. Here the story evolves from the life of a penitent prostitute into the story of a saint. Christ triumphs over the King of Flesh, displacing his worldly amalgamation of ambergris, galingale, and clary with powerful musk, precious incense, cinnamon, and English gillyflower. Unlike Lady Lechery's glittery scented amorousness that belies her carnal touch, the Virgin's scent is sweet, smoky, and pure, blending religious incense most likely encountered in medieval cathedrals with expensive cinnabar and English carnations, which were both known for their medicinal properties. The resurrection transforms the Magdalene's intimate relationship with Christ's body. As the play's famous "Noli me tangere" scene makes clear, Mary Magdalene is no longer the most proximate character to Christ. As he explains to her in the garden scene, her perfumed balms must transform into the sweet smell of virtue, emphasizing a sensory shift from earthly touch to divine smell.

The Magdalene's reward for her penitent life of good works and for her conversion of the "land of Marcyl" is a life of seclusion in the wilderness. Her eschewal of worldly goods emphasizes her saintly transformation. Her vow reaches the heavens as a sweetly scented prayer (that mimics the liturgical practices of censing). Christ exclaims, "O, the sweetnesse of prayers sent onto me / From my wel-belovyd frynd wythowt varyons!" He orders his angels to feed her with celestial sustenance. The angels abide, noting that Christ is a "redolent rose, that of a vergyn sprong!" The play's final theatrical event stages her death, as her soul is escorted into heaven by the angels and her body left to the care of the hermit. The Magdalene's scents are transformed from an earthly materiality into the essence of virtue and prayer. Although no clerk can describe the Virgin's "joy," medieval theater can stage the Magdalene's by invoking both liturgical censing practices and their dispensed scent, conflating religious smells with worldly ones.

Such theatrical representations demonstrate how scents were increasingly used as triggers for understanding religious sense. Tales like the Magdalene's or Richard Wyche's demonstrate that the smell of God was often conflated with worldly perfumes, making interpretations of sensation a fraught component of religious faith. Given new olfactory encounters with perfumes—in worldly markets as well as in the space of the cathedral—sense was increasingly more important then scents in marking the presence of the divine. Conflicts of faith staged new olfactory contexts—and meanings—of perfumed encounters with the divine. When read against the early reformation struggles, olfaction emerges as a troubled threshold of perception, and perfumes as fraught tools of conversion. Thus, the sweet scent of incense does not disappear from England with the Protestant revolution; rather, it is used to dramatize conversion of religious others, particularly Jews, Muslims, even whorish women like the sinful Magdalene who offer their own palpable threat to Christian religious faith. Staged as an important symbol of Christian abstraction, perfume represents the monolithic transformative power of Christianity within a theater of conversion. Yet, as the next chapter demonstrates, such abstraction paved the way for a radical reconfiguration of the body's symbolic presence. Rather than mark the presence of the divine, Henry VIII would use perfume to stage his royal power as unique and embodied, using the very tools Catholic incense was defined against: imported aromatics.

CHAPTER TWO

Casting Selves

Rosewater, Casting Bottles, Court

According to Edward Hall's *The Union of Two Noble and Illustre Famelies of Lancastre and Yorke* (1548), 1522 was a particularly trying year for King Henry VIII, though you would not know it from the staged courtly pleasures held that year. In January, a "great pestilence and death" ravaged London "and other places of the realme," from which, Hall emphasizes, "many noble capitaines died."[1] This must have been especially disturbing to the king, since the French attacked English ships "on euery coast of the sea."[2] As Hall concludes, "wheresoeuer the kyng roade his poore subjects came with lamentacions and cryes."[3] By the end of the month, the king readied his ships for war.

Disease, death, even the threat of war, however, could not stop the court's revelries or the king's pleasure in them (at least according to Hall). In February, the chronicle announces that Pope Leo X declared Henry VIII the defender of the Christian faith, for his noble person was "so formed and figured in shape and stature with force and pulchritude." Such a form must be divine, signifying the "pleasure" that "God wrought" in his shape. God was not the only one who took pleasure in such a form; in March, the chronicle notes, Anne Boleyn arrived at court. Her debut was dramatic; Boleyn was immediately cast in an elaborate masque, *The Assault on the Castle of Virtue,* held in honor of the imperial ambassador. Hall's chronicle collapses these two events perhaps to emphasize the historic irony. Foxe's *Acts and Monuments,* for example, notes that Pope Leo's papal bull was dated October 5, 1521. The king embodies Catholic faith, but, as Hall suggests through his detailed description of the king's assault on Anne's virtue, he celebrates a bodily pleasure with equal "force and pulchritude."

The sumptuous costumes and special effects of the masque celebrated sensuous pleasure. Anne, cast as Perseverance, defended the titular castle of virtue along with seven other ladies, each dressed in a Milanese gown of white satin, wearing bonnets of gold, embroidered with their allegorical part: Beauty, Honor, Kindness, Bountie, Constance, Mercy, and Pity. Banners of hearts, rent or given freely, flew on each tower. Below, in the battlements, "tired like to women of Inde," the female personas of Danger, Disdain, Jealousy, Unkindness, Scorne, Malbouche (slander), and Strangeness lay in wait. The lords besieged the castle with oranges, dates, and "other fruits of pleasure" while the ladies defended themselves with rosewater.[4]

Translated in modern imagination, such details of the staging disappear; the masque is a verbal and visual feast of pleasure. The Showtime original series *The Tudors,* for example, stages the masque's multisensorial exoticism as merely visual eroticism.[5] There are no Indian women, confectionary weapons, or aromatic defenses. There are only campy theatrics: the Lords attack with fake swords, while the women wait in the battlements, scantily dressed in black, verbally rebuking the lords' innuendos. Anne, cast as one of two busty virtues, silently showers multicolored confetti on her assailant. When Henry captures her, they lock eyes and arms.

Hall's description, however, suggests something quite different. The symbolism of war becomes a tool of erotic play, emphasizing instead chaotic, synesthetic sensation. Against the peal of the cannon, Ardent Desire, bright gold flames burnished on his crimson satin robes, leads the attack. Flying dates and oranges, an expensive and rare gustatory treat, bomb the castle. Retaliating with fruit confit, the virtues mount their defense. Rosewater, a substitute for hot oil, defends the castle, drenching the "Indian" women below and soaking into the deep indigo dye of their gowns.[6] Lady Scorn, however, holds her ground, emphasizing that female coldness must yield to the flames of ardent male desire. The masque concludes with the lords taking their ladies "prisoners."

Amid Hall's descriptions of war, plague, and politics, such details seem frivolous. It is perhaps understandable that the Showtime series opted for modern erotic intrigue rather than what has been described as Tudor erotic frivolity. Yet most of Hall's chronicle of 1522 focuses on the sumptuous details of the masque, rather than the events leading to war. He notes that the cardinal gives the king much commendation before departing for Flanders, which suggests that while the masque's exotic details were part of its eroticism, they were also a key part of its political meaning. The rosewater used to defend the castle, the fruity projectiles, and the silks and dyes that attire its players attest to the scope of foreign commodities within English culture. The irony is profound; Henry solidifies his role as defender of the faith by staging his virility with exotic luxury goods. By the end of the decade, one of these luxury goods in particular would help him redefine his kingly power, embodying a different kind of pulchritude: rose perfumes.

The Assault on the Castle of Virtue, critics argue, was fodder for many masques in later Tudor drama. Shakespeare's *Romeo and Juliet*, *Love's Labour's Lost*, *All Is True*, and *Two Noble Kinsman* all contain echoes of Henry and Anne's erotic revelry. Though many cite its influence, few comment on the use of exotic stage props in the production. As the examples of the Wakefield *Offering of the Magi* and the Digby *Mary Magdalene* make clear, Henry's masque was not the first to involve aromatic stage properties. Nor was it the first theatrical use of rosewater; as early as 1464, rosewater "fumigations" were used to protect Elizabeth Woodville, Edward IV's queen, from the stench of London crowds.[7] Revels staged in Elizabeth I's court used rosewater for a similar effect. Robert Moorer, a royal apothecary, provided quarts of rosewater and pints of spike-water to create aromatic stage props of ice and hail in the *Masque of Janus*, presented to her majesty in 1572, and *The Masque of the Wildman*, performed at court in December 1573. What was unique in *The Assault on the Castle of Virtue*, however, is the use of rosewater in staging masculine, kingly "triumph" over exoticized, feminine virtue, particularly given the masque's staging of erotic "Indian-ness." Unlike the damask silks dyed with Indian indigo blue dyes, the rosewater used in Henry VIII's production was most likely distilled in England from damask roses.

The use of rosewater in Henry's masque celebrated a profound agricultural and technological happenstance: the domestication of the damask rose in England coincided with the arrival of the art of distillation, enabling domestic production of rose absolute. Damask roses are synonymous with the art of perfumery; though there are many aromatic species of roses, only two strains can be distilled into an essential oil. And, as Francis Bacon noted in his influential essay "Of Gardens" (1625), the perfume of the damask rose often resides in morning dew, rather than the petals themselves, almost requiring the art of steam distillation to release it: "Roses, damask and red, are fast flowers of their smells; so that you may walk by a whole row of them, and find nothing of their sweetness; yea, though it be in a morning's dew."[8] Whereas other floral petals yield their scents immediately, damask roses are unique in that they often smell better after distillation. Erasmus's *Colloquies* (1523) makes this last point explicitly, when Pamphilus explains to Maria that the rose, as a symbol of virginity, contains both visual and olfactory "delights" and that, in his opinion, "the rose that withers in a man's hand, delighting his eyes and nostrils . . . is luckier than one that grows old on a bush."[9]

The domestic production of rose perfumes previously associated with foreign places like Damascus had profound political ramifications in England. Though they may seem ephemeral and fleeting, a trivial fashion lost to history, rose perfumes enabled a new kind of kingship to emerge, one that would define Tudor power in the sixteenth century. Rose perfumes, I argue, enabled both Henry VIII and Elizabeth I to yoke the divine right of kingship to Protestant definitions of royal power, defined by a material, bodily embodiment. The doubled Tudor rose was a powerful sign of English peace in the early sixteenth century, unifying the white and red roses that symbolized the warring houses of York and Lancaster a century before. The damask rose was valued for its scent as well as its color. With double the number of petals of wild English roses, each blushing with color at its ends, it visually embodied the dynastic Tudor rose. And, when distilled, it added an olfactory dimension to the political and spiritual understanding of kingship in the period. As Henry VIII undoubtedly realized, its perfume enabled him to perform Tudor dynastic power. Such an olfactory breakthrough greatly amplified his regal presence at court, just as incense defined the invisible power of transubstantiation in the Catholic mass.

The arrival of the damask rose in England was thus timely: its concentrated essence underscored the power of Henry VIII's two kingly bodies. Infusing his physical body with fragrant scents previously associated with divinity, rose perfumes enabled him to heighten the olfactory experience of kingship for those in his wake and to provide an olfactory reminder of his presence after he left. As *The Assault on the Castle of Virtue* suggests, Renaissance self-fashioning in England could be decidedly aromatic. Perfumes were key tools of transformation in Henry's court just as they were in the Catholic mass, defining control over distinct spaces. Whereas medieval incense captured all that could not be embodied through earthly materialism, Henry's rose perfumes allowed him to perform a perfect amalgamation of his earthly and kingly bodies.

Tudor Roses and the Smell of the King's Two Bodies

Henry VIII's early reforms dramatically altered the way perfume was produced and consumed in England. Tudor gardens were already markedly different from their medieval counterparts. Drawings from the mid-sixteenth century document royal gardens "profusely adorned" with "heraldic ornament"; these were public spaces designed to impress visitors rather than private sanctuaries designed to provide solace or relaxation.[10] The damask rose provided a natural counterpoint to such decorations, its scent amplifying the royal garden's visual splendor, extending the king's courtly presence into outdoor space, even if he was not physically present.

The dissolution of the monasteries greatly increased the number of royal gardens in England. When royal gardeners transformed monastic retreats into public spaces, they also altered England's perfuming practices. No longer controlled by the episcopacy, perfume production became a domestic affair. English perfumes were no longer just incense-based scents or perfumes meant to be burnt; more and more, they were elaborately scented waters, blended from distilled oils, spices, and flowers. For example, when Henry VIII took control of Cardinal Wolsey's garden at Hampton Court in 1529, he greatly expanded its pleasure gardens, adding an orangery, a bowling alley, a tilting yard, a sweet garden, and a rose garden.[11] Along with primroses, violets, gillyflowers, sweet william, and mint, the king ordered over a hundred rosebushes, including damask rosebushes. Such spatial reconfigurations suggest how olfactory pleasure and erotic pleasure were combined. Some garden historians claim that the king and Anne Boleyn often met in the rose garden for erotic dalliances. That same year, the king's personal apothecary distilled twenty-nine different essential oils and gave casting bottles as gifts to his twelve sexual partners during the celebration of Twelfth Night. Thus, by 1529, perfumes were routinely part of royal gifting cultures, including adulterous ones, demonstrating a reconfiguration of perfumed scents from religious registers toward erotic ones. By the end of the sixteenth century, rose attar was an important part of staging an embodied self.

The arrival of the damask rose in England was a key component of this shift. It is said that the first damask rose arrived in England in 1522, given as a gift to the king from Dr. Thomas Linacre. Linacre, founder of the Royal College of Physicians and royal physician to Henry VII and Henry VIII, had recently returned from Italy, where he had been studying medicine at Padua. A great lover of nature, Linacre was inspired by the technological, artistic, and agricultural advances of Renaissance Italy. According to David Hume, he returned to England in 1522 with the damask rose and perhaps the means to extract its scent.[12] In this telling, domestication of the damask rose stands at the epicenter of the English Renaissance. It is an important marker of early modernity in England, including the king's reformation, the evolution of a body-centered individualism, the arrival of scientific breakthroughs from southern Europe, and increased secular consumption of luxury goods. Perfumes are routinely invoked as part of the divide between medieval and early modern Europe; Jacob Burckhardt, for example, famously uses the perfumed scents emanating from the preserved corpse of a young Roman girl, discovered in 1485 during the excavation of the Convent of Santa Maria Nuova, to emphasize the civilizing power of the Renaissance.[13] In both instances, bodily scents are conflated with perfume, illustrating a distinct emphasis on Renaissance individualism.

Although the damask rose was hardly a new floral specimen in the sixteenth century, its domestication in England was a profound agricultural achievement. Noted for its scent, *rosa damascena* had been bred in eastern Europe since the fifth century BCE, and its aromatic petals had long been used as offerings, first to Aphrodite and then to Venus. Believed to be a hybrid of the western *rosa gallica,* so named by Linnaeus for its Gallic origins, and the *rosa phoenicia,* which grew all over the southwestern parts of the Ottoman empire, or modern Turkey, *rosa damascena* represented the best attributes of both. Like *rosa gallica,* whose bright red petals retained their scent even after they withered, and *rosa phoenicia,* whose pale white petals were valued for their perfume, *rosa damascena* offered a visual and an aromatic pleasure that lingered.

Damask roses, as their name suggests, were valued for their "outlandishness," an early modern term that emphasized foreign specimens. First bred in Syria, the rose was introduced to continental Europe by crusaders who brought the aromatic specimens home with them.[14] The damask rose was thus unlike any other rose in sixteenth-century England and was practically made for perfuming: its shrubs were hardy, its floral scent was powerful, and, prior to the nineteenth century, it was the only remontant rose in England, blooming twice, once in spring and again in August.[15] That its "outlandish" scent became conflated with English roses, English pleasure gardens, and English perfumes demonstrates the importance of olfaction in crafting national identity. And, because the scent of the damask rose was distilled into a perfume meant to be worn on the body, rather then dispersed into space, it linked national identity with more personal notions of identity.

Distilled perfumes intensified such associations. Shakespeare's Sonnet 54, for example, wonders how olfaction intensifies visual aesthetic pleasure in beauty, metaphorically linking rose perfume with a lover's essence.[16] Comparing his lover's beauty to that of the damask rose, whose beauty seems more "beauteous" because of the "sweet odour which doth in it live," the poetic narrator of Sonnet 54 emphasizes the role of perception in aesthetic desire: the flower's sweet ornament—its scent—intensifies its beauty (lines 1, 3). Unlike the scentless canker roses (native to England) that live "unwooed and unrespected" and "die to themselves . . . / Sweet roses do not so; of their sweet deaths are sweetest odours made" (lines 10–12). Distillation of the sweet roses into rose absolute is a metaphor for both erotic and poetic reverence. Beauty may wither and die, but its "beauteuous" effects can be bottled for later pleasure. The poet's verse "distills" the youth's "truth" into something that transcends space and time, just like perfume captures the scent of the rose, normally released only through the "wanton" play of summer winds on the flower's bud. Does the poet's verse embellish the truth of beauty, just as perfume acts as a cosmetic? Or is poetry itself a kind of wanton play, the perfume distillate just as aesthetically pleasing as smelling a rose firsthand? The poem suggests both. That Shakespeare invoked distilled rose perfumes as a metaphor for both sexual consummation and the poetic process demonstrates how embodied olfactory pleasure and its links to eroticism troubled emerging notions of Renaissance selfhood. Perfume is imagined both as a "beauteuous" ornament—a cosmetic addition to or correction of the body—but also as the "truth" of that beauty, a portable distillate that enables others to perceive that "which doth in it live."

Perfume was not the only commodity that participated in this process; however, its invisible, ephemeral quality troubled the relationship between embodiment and olfaction. Where did the body end and one's environment begin? Which influenced the other more fully? Was passion intrinsic or extrinsic, especially when it was inspired by proximate sensation? Unlike the paradoxes of incense, which conflated and confused what was imagined as a divine olfactory pleasure with the earthly materials that produced it, rose attar—whose perfume was stronger than its original, floral form—emphasized the human body, providing a way of extending it into space and of representing that extension as an intrinsic, "essential" component of the self. Rose attar combined the symbolic power of the Tudor rose with the exotic, intoxicating foreignness of the transplanted Damascus rose into a perfect distillate of erotic power.

The art of distillation, fueled by the discovery and widespread use of ethyl alcohol, transformed the production of perfume in the late Middle Ages. Though distillation was hardly a new discovery—a terracotta alembic vessel dating from 3000 BCE was recently discovered in Pakistan—its genealogy is complex. Popular histories of chemistry, like those of perfume, link its discovery to ancient Egypt. By the early fifteenth century, Zosimus of Panopolis attributed its discovery to Maria "the Jewess," an early practitioner of the "sacred" arts of metallurgy, who lived in Hellenistic Egypt (most likely Alexandria) in the third century CE.[17] Further techniques were described and perfected by Jābir ibn Hayyān, in the seventh century CE, and by the Persian physician, Ibn Sinā, or Avicenna, in the eleventh century.[18]

Distillation's Arabic and Jewish origins within Renaissance European genealogies highlighted the outlandishness and mysterious power of its two most prominent applications: producing distilled spirits and liquid perfumes. Part of the allure of these commodities was their exotic origins; domestic production of both was rare in England prior to the sixteenth century. Most were imported into England by Spanish and Portuguese traders and were expensive to buy; commodities such as Canary sack, a fortified wine from the Canary Islands, and Moroccan rosewater, distilled from damask roses, were increasingly associated with "dark" religious reform.[19] Describing the city of Fez, for example, Leo Africanus provides a brief gloss on the history of alchemy, noting that the city is full alchemists, "who are most base fellows, and contaminate themselves with the steam of Sulphur, and other stinking smels."[20] Africanus's text, written in the mid-sixteenth century, emphasizes that English distillation is superior. Jābir ibn Hayyān, misidentified by Africanus as "Geber," is given particular prominence, though it is noted that "being a Greeke borne," he "is said to haue renounced his owne religion" and that "his works and all his precepts are full of allegories or darke borrowed speeches."[21] In his *Treatise of Newe India* (1553), Richard Eden also noted that Arabian priests washed the devil with rosewater.[22]

By the end of the sixteenth century, however, rosewater was retroactively imagined as a fully English commodity. Nashe's *Unfortunate Traveller,* a fictional account of Jacke Wilton, a page in Henry VIII's court, mocks university orators whose beards were "plentiful besprinkled with rosewater."[23] Likewise, an early seventeenth-century book of housewifery published a receipt titled "King Henry the eighth his perfume," calling for six spoonfuls of "Rose-water," along with grains of musk, ambergris, civet, and cloves.[24]

Henry VIII's aromatic embodiment of Tudor power was a key part of this process of making rosewater English. Rosewater was part of the English reformation, surpassing Catholic incense as a sign of power. It imbued the royal body (and, later, bodies in general) with a unique and alluring scent associated with the concentrating processes of distillation. Using rosewater, the absolute essence of the rose itself, as a subtle olfactory reminder of his absolute royal power, the king embodied a new kind of self, one whose identity relied on olfaction. Rose perfumes redefined the body and its boundaries to include material smells that were extrinsic to it. That the scent of damask could perform English royalty demonstrates that the body's material history is fluid and fungible.

The history of domestic rose attar in England is thus an important example of how bodily materiality changes due to technological advancements and cultural shifts. Linked to the practice of alchemy, with its emphasis on the essential substance of matter, the art of distillation was one of transformation. Alcohol, known as aqua vitae, or the ardent spirits of the first distillation from wine (*aqua gardens*), was believed to be the quintessence of life, an invisible substance more subtle and pure than earth, air, fire, and water.[25] Alcohol was a pure contradiction, a liquid that burned, and ethyl alcohol, early chemists soon discovered, was a perfect conduit for spreading the aromatic scent of rose absolute. Alcohol, a product of distillation, could dissolve materials that would not dissolve in ordinary water, namely other essential oils produced from distillation. When applied to skin, it evaporated quickly, releasing those aromatics into the air. Prior to the discovery of ethanol alcohol, other oils were used to convey the aromatics. Theophrastus advised using sesame oil, due to its viscid nature, though most perfumes were smokes or balms.[26] Rose attar was thus produced through the most advanced medieval technology, a refinement of elemental matter into an invisible, paradoxical substance that smelled stronger and sweeter than its original form.

Though many flowers contain essential oils, including a host of aromatic roses, only the scent of damask roses can be captured through the process of steam distillation.[27] The sole other species of rose routinely used in modern perfumery is the cabbage rose of France. Known also as the *rosa centifoila,* or hundred-leaved rose, this flower can yield sweet-scented absolute oil. However, it requires additional steps: a solvent is used to extract the aromatics from the flower petals, which are then mixed with alcohol and cooled, then finally heated again until the alcohol evaporates. Steam distillation remains the simplest of distillation processes, requiring only a bain-marie or a double boiler, making it something that could be done at home. For this reason, damask roses were associated with rose attar until the advent of the modern perfuming industry in France in the eighteenth century.

By the mid-sixteenth century, more and more books on distillation were being printed, aimed at common artisans, alchemists, barbers, apothecaries, and housewives. Though the process is called simple, it is not easy; it may take upward of five hundred pounds of roses to produce one pound of rosewater, or hydrosol, and much less essential oil.[28] For best results, the petals should be picked first thing in the morning and distilled immediately for a soft, sweet scent.[29] Each floral essence requires specific knowledge; lavender, for example, should be distilled very briefly over very high heat, while vetiver should be distilled for at least twenty-four hours over low heat. By the mid-seventeenth century, recipes would specifically call for the "Artificial Preparation of Damask-Roses, for Smell."[30]

Attar of roses is a powerful scent. Theophrastus described it as obscuring all others whereas Galen notes its cooling effects. The rose oil Galen describes was most likely produced from damask roses since he observes that the best rose perfumes are produced in Phaselis, an ancient harbor city located on the Pamphylian gulf in the province of Antalya in modern Turkey.[31] Likewise, in one of his seven books of medicine, Paulus Aegineta, a Greek physician working in Byzantine Alexandria, discusses the "powers" of "simple medicines" and the difference between things perceptible and things imperceptible.[32] To illustrate this, he describes the aromatic properties of rose oil as it is broken down into simple medicines: "[L]et such an order be laid down to render clearer the course of instruction, rose oil or the rose itself being placed in the first order of cooling things; the juice of roses in the second, and in the third and fourth those things which are extremely cold, such as cicuta, meconium, mandragora, and hyoscyamus."[33] Within Galenic systems of health, roses were noted for being cold in the first degree and dry in the second, but by the mid-sixteenth century, English medical texts emphasize that damask roses are lighter than other rose specimens and more astringent.[34] And in his sixteenth-century herbal, Leonhart Fuchs, quoting Plutarch, writes that roses were called Rhodon because they "sent forth a great outpouring of perfume."[35]

Rose oil was not associated with royal perfumes until the sixteenth century.[36] Whereas perfumes were an aromatic luxury enjoyed by ancient royalty, floral essences were rarely part of their composition. Pliny, for example, describes perfumes prepared for the kings of Parthia that were simply titled "the royal perfume," composed of oil of benjamin. None involved rose oil, which Pliny values instead as a medicinal treatment for headaches.[37] And when such essences were not associated with health, they were associated with godliness. Given the intense perfume of the damask rose, it is not surprising that John of Damascus describes the imperceptible power of God as a "rose tree, a flower, and a sweet fragrance."[38] In medieval European allegories, roses signified a virginal, feminine purity. St. Bonaventure's *Meditations* allegorically links the rose to Mary's virginal innocence, whereas the thirteenth-century French allegory *Roman de la Rose* associates the plucking of a rose with the pursuit of a woman in traditions of courtly love.

Imported rosewater was expensive: a leet book from Coventry recorded in 1456 that a glass of rosewater cost two shillings.[39] Rose attar was even pricier; such a rarified essence was used only in limited ways by the very wealthy. By the early seventeenth century, however, Gervase Markham would advise that every housewife "furnish herself of very good stills, for the distillation of all kinds of water."[40] Such advice documents the profound shift in perfuming practices that occurred across the sixteenth century. The art of distillation enabled English men and women to capture the perfumes of floral specimens like the damask rose and produce their own perfumes.

Prior to the sixteenth century, distillation was concentrated in English monasteries. The earliest distillers in England worked in royal households and monasteries with "stills" most likely producing aqua vitae.[41] Medieval monasteries, equipped with botanical gardens and stills, produced the earliest perfumes in England. A simple still works in the following way: aromatic petals of flowers are soaked in water and placed in a still. The water absorbs the oils from the petals; when the still is heated, steam vaporizes the aromatic oils of the petals, forcing them through a condenser where it once again separates into oil and water. The oil produced is essential, whereas the remaining water solution, which bears trace amounts of the oil along with nutrients from the plants, has a subtle scent and is known as hydrosol.

The widespread production and consumption of perfume in sixteenth-century England were also part of large-scale changes in the practice of medicine. Whereas Galenic practitioners defined health as an endogenous balance of humoral fluids, early chemists and some medicinal practitioners, influenced by the writings of Paracelsus, a Swiss chemist, botanist, and physician, began to argue that health was better achieved by attending to exogamous elemental matter rather than bodily effluvia. The practice of distillation by early alchemists helped support such conclusions, showing that organic matter could be broken down into smaller component parts—liquors, inflammable oils, and solid residue. Some alchemists went so far as to describe the earth itself as a giant alembic vessel from which elemental particles are formed; others concluded that a man could be made in a glass cucurbit from human sperm.[42] Such beliefs suggested that unseen and ephemeral material, like foul and fragrant air, could greatly influence bodily health. Paracelsian ideas of chemistry quickly infiltrated vernacular medical traditions, especially in print. Early treatises on distillation were an important part of this process and were quickly translated from Latin into English for alchemists, apothecaries, and medical practitioners working far from city centers.

The most influential of these texts was Hieronymus Brunschwig's *Vertuose Boke of Distyllation*, translated into England in 1527.[43] Brunschwig defines distillation as a purification process, and a paradoxical one at that. Through it, the gross becomes subtle; the corruptible, incorruptible; and the material, immaterial. "Dystyllyng is none other thynge, but onely a puryfyeng of the grosse from the subtyll, and the subtyll from the grosse / eche seperatly from other / & to the entent that the coruptyble shall be made incorruptyble / and to make the materyall immateryall."[44] Such profound transformations of matter were the stuff of alchemy, though Brunschwig's book seeks to make this process more readily available to doctors and other medicinal practitioners located far from city centers.

Brunschwig is attentive to frugality. His book provides five ways of distilling without cost: his directions call for a three-cornered still and advise readers to draw on the heat of the sun, an oven already employed in baking bread, a pile of horsedung, or, alternatively, to bury it in an anthill. Recipes include waters of marjoram, sorrel, blue violets, endive, fennel, iris, mercury, thistle, and goat's milk, to name just a few. Brunschwig's text, which was written in Germany in the late fifteenth century, does not name the damask rose specifically, outlining instead receipts for waters of "rose alba," "rose eglentyne," "wylde roses" "rede roses," and "whyte roses."[45] Such directions, however, were easily adaptable—"the same waters may be dystylled of what roses that you please." Within Brunschwig's smellscape, the best waters came from red roses, followed by wild roses, white roses, and eglantine roses, in that order.[46]

Konrad Gesner's influential *The Treasure of Euonymus* (1559) and *The Newe Jewell of Health* (1576) emphasize that chemical distillation produces effective medicines. Though Gesner's works do not specifically refer to damask roses, the production of rosewater is a key pedagogical device throughout both texts, suggesting that it as one of the first processes would-be distillers would learn.[47] And as early as 1545, damask roses were prescribed to sailors and "curious" passengers about to embark on sea voyages in order to purge their "yll humors," noting that these roses were now integral to health and well-being.[48] Unlike other foreign floral specimens, the damask rose was imagined as an English tool of travel rather a symbol of it. The scent of the rose performed an act of spatial hybridity, translating Damascus, Alexandrina, and Persica into English space through the art of distillation and alchemical technology. Its scent thus enabled more and more potential English travelers to engage with other environments and their own more fully. The rose retains its outlandish status, but only as a marker of English mastery; its value is its aromatic properties, distilled to sweeten English bodies and selves.

Distillation thus enabled new kinds of scent practices to emerge, specifically the use of liquid perfumes on the body, dispensed from small, intricate casting bottles. Casting bottles concentrate essence on the body itself, rather than in a space or on clothing. Compare, for example, two sixteenth-century scent objects linked to Henry VIII. The first is the golden rose, a consecrated object conferred by the pope on kings of Europe to recognize their Christian sovereignty. Few still exist, though the Musée National du Moyen Age has a rare fourteenth-century one on display. Meant to signify the divine power of kingship through both the object's visual splendor and its aromatic scent, the golden rose was conferred on Henry VIII twice before his separation with the Church of Rome. A practice dating from at least the twelfth century, the gifting of this holy object linked the monarch's sovereignty to Christ's transcendence over human death on Easter Sunday. On the fourth Sunday of Lent, after being used in a host of processionals, the pope consecrated the golden rose by perfuming it with holy balms and incense and placing on the altar during Mass. Only then was it sent off as a royal gift.[49] In 1510, Pope Julius II consecrated the golden rose, anointed it with musk and balsam to symbolize the sweet smell of Christ, and presented it to Henry VIII. Pope Clement VIII sent a second one in 1524.

The second scent object is a small casting bottle on display at the Victoria and Albert Museum in London. Whereas the golden rose defines the king's divine power by aromatically linking it to incense, the bottle demonstrates how perfume might define the power of the king's body in other ways. Like the golden rose, the casting bottle is a rare artifact.[50] As the display notes, it was most likely refitted twice, from Egyptian rock crystal into a Catholic religious reliquary in the tenth century and then into a casting bottle in the sixteenth. Mounted in silver gilt and embellished with acanthus leaves and a chain, this type of bottle was a popular gift in Henry's court. Its small size is misleading. Filled with powerful floral essential oils and worn on a chain, it enabled its wearer to cast its perfumes repeatedly on his or her body, imbuing the space of court with its scent. Repurposed from medieval religious uses and filled with rose attar, the casting bottle was a powerful tool of self-presentation.

Together, these two objects demonstrate a profound change in the meaning of roses and their scent, which shifted from Catholic, iconographic visual representations, where the rose is a symbol of faith and purity that transcended earthly things like human bodily smell, toward an active, erotic engagement with the smells of embodiment.[51] The odor of sanctity becomes the odor of bodies. With the use of perfumes, olfaction became synonymous with sexual pleasure, and casting bottles became tools of eroticism, as later sixteenth-century representations of gifting rituals, specifically those involving rose perfume and casting bottles, make clear.

Bullein's *Bulwarke of Defence* (1572), for example, includes a "Booke of Simples," in which he describes how to make a "water of the casting glass," along with two other recipes for sweet waters; he also mentions perfumed gloves scented with "sweete rose, or Damaske water."[52] Such sweetly scented gloves make excellent gifts, he notes, for "cytizens that dwell in close, corrupted foule ayre." Link-

Minuccio Jacobi da Siena, Golden Rose (Papal Rose), ordered by Pope John XXII in 1330. Gold and colored glass. From the Treasure of the Cathedral in Basel, produced in Avignon. A gift of a golden rose, blessed with incense by the pope, symbolized Christ's love and Passion and celebrated the recipient's devotion and service to the Church. Musee national du Moyen Age—Thermes de Cluny, Paris, France. Réunion des Musées Nationaux / Art Resource, NY

Casting bottle (circa 1540–50). Carved rock crystal set in silver gilt. Made in Egypt in the ninth century and mounted in England in the mid-sixteenth century. Photo © Victoria and Albert Museum, London

ing this perfuming practice to "noble Princes, Lords, Ladyes, and gentle folkes," Bullein argues (in the inverse) that the sense of smell is a bodily pleasure and perfume a tool of godliness and a noble practice that should inspire average citizens, for "God gyue them grace to know it to bee hys gyft, to pleasure the Sense of smellyng, and to defend euyll ayres, and not to moue, or to be an instrument of Ungodlynes."[53] And Gascoigne's libelous and immoral *Adventures of Mister FJ*, published in 1573 and quickly suppressed in 1574, also shows that liquid perfumes were tools of bodily pleasure used by the nobility. Gascoigne's romance narrates the adulterous relationship between a married woman, Dame Eleanor, and a nobleman, FJ. Dame Eleanor entices FJ to her bed by casting damask rose-water onto an already sweetly scented pillow (from her "odiferous chest") and "bedewed his temples with water which she had ready in a casting bottle of Gold."[54]

The popularity of casting bottles as fashionable gifts influenced phenomenological descriptions of sex. For example, Robert Dallington's 1592 translation of Francesco Colonna's *Hypnerotomachia Poliphili (The Strife of Love in a Dreame)* describes olfactory pleasure as perfumed. Before he is brought to Queene Eutherillida, Poliphilus, the main character, takes an erotic bath with five nymphs, meant to represent the five senses, "to refresh and delight" his own senses. The nymphs' attributes describe multisensorial luxury, and Dallington translates nymph Offressi's attribute as a box of "sweete perfumes" and "casting bottels of golde and precious stone" filled with "pretious Lyquor."[55] Likewise, John Florio's 1598 English–Italian dictionary, *A Worlde of Wordes,* includes two translations for those hoping to purchase casting bottles on their trips to Italy: *Acanino,* defined as an "eawer, a casting bottle, a glasse to keep sweet waters in," and *Oricanno,* "a casting bottle to cast sweet waters with."[56]

By the early 1600s, the gifting of perfume was explicitly linked to sex. In Nathan Field's *A Woman's a Weather-cocke* (1610), a servant's pun links "casting" to both the cause and effect of pregnancy. Commenting on his pregnant mistress's morning sickness, he remarks, "what a casting shee keepes, marrie my Comfort is, wee shall heare by and by, who has guien her the Casting Bottle."[57] And in Henry Chettle's *Tragedy of Hoffman* (1602), Prince Jerome, who fears that his previous tilting "hath brought" him "into a strange sauor," so much so that he "appeares not like a Prince," calls for his casting bottle and commands his servant to "sprinkle . . . sprinkle."[58] Shakespeare's Autolycus in *The Winter's Tale* commands "lads to give their dears" a host of aromatic gifts including "[g]loves as sweet as damask roses; / Masks for faces and for noses; / Bugle bracelet, necklace amber, / Perfume for a lady's chamber" (4.4.214–25).[59] Likewise, Ophelia "redelivers" Hamlet's amorous "remembrances" because they had lost their "perfume" (3.1.101–2), and in *The Taming of the Shrew* Gremio sends Bianca papers "very well perfume'd, / For she is sweeter than perfumer itself" (1.2.144–45). By the mid-seventeenth century, damask roses are imagined as both the cause and cure of desire, their aromatic scent arousing and purging the body of pleasure.[60]

Perfume was so associated with eroticism and casting bottles with love tokens that Shakespeare could pun on both in his sonnets. As Richard Halpern has argued, the metaphor of the perfume and perfume bottle in Shakespeare's sonnets enabled "a curious material demonstration, even before the fact, that sexual desire can be sublimated into art."[61] Whereas Halpern focuses on the distillation of perfumery into metaphorical abstraction, such metaphors also provide a way of understanding perfume's "curious" materiality. Though Sonnet 5 does not name the flower whose visual and odiferous beauty becomes a "summer distillation left / a liquid prisoner pent in walls of glass," Sonnets 99 and 130 suggest that it is most likely a damask rose (line 10). Sonnet 99 chides red and white roses for "stealing" the lover's blushing beauty and "pink" roses for stealing both visual beauty and the perfume of the lover's breath.[62] That the sonnet links white and red roses to color and pink roses to color and smell suggests that the pink roses were damask roses. Furthermore, Sonnet 130 famously calls the damask rose by name: the poet has "seen roses damask'd, red and white, / But no such roses see I in her cheeks; And in some perfumes is there more delight / Than in the breath that from my mistress reeks" (lines 5–8). Once again, the poem links red and white roses with blushing beauty and "damask'd" roses with perfumed breath.

Read together, Shakespeare's sonnets gesture toward the ubiquitous association between damask roses, perfume, and eroticism in the late sixteenth century, so much so that eroticism in his *Venus and Adonis* is imagined as olfactory. Venus declares her love of Adonis to be so strong that even if she were bereft of all her senses except smell "and that [she] could not see, nor hear, nor touch," her love would increase by the smell of his breath alone, distilled from his face: "for from the stillitory of thy face excelling / Comes breath perfumed, that breedeth love by smelling" (lines 433–44).[63] Because smell is a key part of sexual attraction, the art of distillation is metaphorically imagined as a process through which a lover's face is an alembic still that "breedeth love by smelling." Such breeding puns on earlier metaphors of poetic reproduction: poetry is like a perfume, a distillate of material beauty, captured for others' pleasure and saved from the ravages of time.

Such metaphoric uses of distillation indicate that the art of perfumery enabled a profound alchemical transformation to occur, a transformation in which the spirit of the substance is captured and concentrated. As a tool of eroticism, distillation also suggested bodily smells. Pleasant scents like perfume distilled from damask roses could function as a synecdoche for other kinds of olfactory pleasures, namely bodily ones, such as pleasure in a lover's breath or sweat. The art of distillation enabled the damask rose and its exotic scent to become domesticated and eroticized, and the first step in both processes was the profound association of rose attar with the body of Henry VIII.

Whereas the golden rose conferred on Henry was imbued with incense, emphasizing the divinity of his kingly body, the small casting bottles gifted to

Henry's mistresses contained a very different kind of scent, at once both powerful and intimate. In such gifting practices, rose attar was not the scent of godliness but rather of kingliness. Its scent was meant to amplify the king's bodily presence. Casting bottles were expensive luxury items, as were the perfumes contained within. The casting bottles worn on his mistresses' waists showed their physical connection to the king while the scent of the damask roses indicated a more intimate connection to the king.

Given the ways perfume is invoked in late-sixteenth-century love poetry, one might imagine that the king gifted perfume for his own olfactory pleasure, as a gift that enabled him to enjoy smelling it on a lover's body. Read against Tudor heraldic symbolism, however, the gift of rose perfume suggests something quite different: rose attar was the potent scent of the king himself, his substance distilled, concentrated, and made portable. As such, its scent emphasized an intimate connection with the king's two bodies—the eroticism of his body natural and the dynastic power of his body politic. Its scent thus encapsulated a distinctive break with both medieval political philosophy and the Catholic Church.

The Tudor rose provides one way of understanding such regal self-fashioning. Emblazoned on everything from portable locks to dog collars, the Tudor rose visually marked the property of the king. Likewise, attar of roses would have aromatically linked the wearer to the king. Henry VIII played with such a motif as early as 1526, issuing debased English crowns, worth five shillings each, with the motto *Rutilans Rosa Sine Spina*, "a dazzling rose without a thorn," a phrase refitted from the Virgin Mary.[64] The virgin's rose, however, was the heath rose of Jericho, a small herb with small white flowers that bloom at Christmastime. As Tudor herbalist John Gerard points out in his *Herball or General Historie of Plants* (1570), however, "the coiner spoiled the name in the mint, for of all plants that have bin written of, there is not any more unlike unto the rose" in "shape," nature," or "faculty."[65] Using a numismatic etymology, Gerard perhaps knowingly puns on this Tudor motto. For Elizabeth I, too, would adopt this mantra as her own, minting pennies emblazoned with her portrait and this phrase, yoking religious virginity and her father's monarchic strength through the symbol of the rose.

Like her father, Elizabeth I utilized the rose's olfactory and erotic significance; her gardens at Hastings grew elaborate rose bushes, though with an important distinction. These roses, and their powerful scent, symbolized royal chastity. In June 1581, Elizabeth staged her own courtly masque in honor of the arrival of the French embassy in London. In the midst of negotiations of a possible alliance with the French, one hopefully made without a marriage to Hercule-François, duc d'Anjou, the queen staged "The Four Foster Children of Desire" at Whitsun.

As in *The Assault on the Castle of Virtue,* damask roses were a key part of the production. Using roses to defend her castle of virtue more propitiously than her mother, Anne Boleyn, had, the queen's masque made clear that there would be no marriage. In the performance, the fortress of perfect beauty (allegorically representing the queen) is besieged by four knights of desire, the "children" of Desire, representing Anjou. A peal of cannons, ejaculating "sweet powders" and "sweet "waters," signaled the attack. Both were "very odiferous and pleasant," while footmen "threw flowers and such fancies against the walls, with all such devices as might seem fit shot for Desire."[66] That the queen was able to withstand such an amorous assault stemmed from her uniquely Tudor constitution. As she later famously declared, her heart, her stomach, and her royal power derived from her dynastic lineage; though her body natural was gendered female, her body politic was not, a point perfectly distilled in the rose powders, waters, and flower petals used in the masque and embodied in the phrase *rosa sine spina* printed on her royal coins.[67]

The Rose Theatre and the Smell of Bankside

Rose attar thus linked early modern English embodiment with an olfactory performance of self. Such invisible, ephemeral performances were central to early modern London's dramatic entertainments that converged, appropriately, in the city's theatrical gardens to the south, namely the Rose Theatre, where Shakespeare and his plays made their London debut. Its balconies enabled plays to stage both domestic interiors and outdoor space; its lanterns introduced a stage effect of night; and its doubled discovery spaces provided a way to convey deep interiority and intimacy, which was key to staging both tombs and bedchambers. They also enabled a host of properties to enter the stage without the effect of curtains. Yet the name Rose Theatre prompts certain questions about its multisensorial realm, both in Bankside and in London: what is the relationship between the spectacles performed there, Renaissance self-fashioning, and the spatial location of the stage? Did the Rose connote other olfactory meanings that were significant for early modern theatergoers and to the bodies and selves performed on its stages?

In his influential study of the "place" of the stage in early modern London, Steven Mullaney argues that early modern theater gained a central place in the imaginative landscape of the city by occupying physical space outside it; the theater "materially" embodied the contradiction between its extrinsic location and its intrinsic cultural relevance. Famously, the external liberties, or outlying areas, of London that the theaters occupied—Holywell, Shoreditch, Bankside—were defined by a sense of openness that inspired lawlessness and provided those identified as others within the city walls with a space to congregate.[68] Marshes, farms, and cemeteries became bear-baiting venues, sinks, stews, and sewers.[69] These spaces, defined in part by their stench, provide rich imaginary *topoi* against which Renaissance theater emerges: the Rose springs forth from squalor. For ten years, it stages perhaps the most important plays in English literary history, defining audience expectations for the burgeoning form. Then it disappears, fading from the city's spatial memory, eclipsed by the Globe Theatre, built nearby in 1599, and the Fortune Theatre, built in 1600 to the north—both possessing names that symbolize the scope (and economic appeal) of early modern drama.

Such an interpretation has its appeal, particularly in the way it links Shakespeare's stage with a popular audience almost from its inception. But its eagerness to situate the theater within a particular time and space imbues it with a material ephemerality: the Rose buds, blooms, and fades from London's landscape. As Mullaney's reading suggests, the difficulty in describing any space is that it is always already a failed project. Early modern writing about London describes a city that, almost immediately after the ink had dried, had already vanished. More recently, critics have approached this space not as one described but rather as one inhabited, focusing on how bodies moved within it and relying on the kinds of local knowledge and microhistories that slowly permeate city dwellers' awareness in crowded, urban life. London in the 1590s, as Ian Munro argues, was "an experiential space" defined by "the visible and tangible presence of more and more *bodies*."[70] In both Mullaney's and Munro's theses, the place of the stage was defined in relationship to those bodies as a breeding ground for licentiousness and rebelliousness that allowed new kinds of artistic representation to emerge.

The Rose Theatre thus connects to Henry VIII's aromatic performance of his kingliness with rose attar through the aesthetic, allegorical, and alimentary meanings of roses, particularly the phenomenological effects a body can have on space.[71] For example, most early modern theatergoers familiar with Bankside would immediately recognize in the name of the Rose a winking reference to both Tudor iconography and the foul and fragrant smells of the area's many stews, including those stocked with fish and those associated with prostitutes.[72] Those who had navigated the area for a long time would also recollect that the Rose Theatre was not the only Rose at that same site. As early as the 1480s, records suggest that there was a brothel called the Little Rose in this location.

Henslowe's 1585 lease also specifically mentions a "Tennement the Called the little rose" on the "east sid" of "the barge the bell and the Cocke," three well-known brothels in the area.[73] That the theater assumed the name of a brothel should not be surprising: the entertainment (and economic success) of the little Rose was, in many ways, a part of licentious entertainment the (big) Rose hoped to embody.

Yet the names of both the big and little Rose also hint at other kinds of spatial and sensory ways of knowing, connecting their entertainments in the 1590s with earlier Tudor history, specifically the pleasure associated with smelling a rose. As early as 1547, records suggest that there were two rose gardens on this site, which was the property of the aristocrats who had leased the land to the brothel's owners; in 1585, the gardens were still there, enabling those traversing the space to experience firsthand the scent of kingly power and pleasure.[74] Certain theatrical entertainments intensified such connections. For example, a visitor to London in 1584 describes the day's entertainments at the Rose Theatre as including bear-, horse-, and bullbaiting; fistfights; food fights; and "a rose," fixed high above the crowd, "being set on fire by a rocket" so that it rained "apples, pears, and rockets" on the spectators below.[75] The day's many festivities culminate with the symbol of Tudor power—and the theater—exploding into multisensorial pyrotechnics.

The rose gardens in Bankside that abutted both the little Rose and the Rose Theatre were odiferous, according to the many early modern sources that point out the juxtaposition between London's horticultural and theatrical "gardens." For example, in the poem on the "whoremonger" in Richard West's *Court of Conscience* (1607), the path to the theater (at least in the city's northern suburbs) leads through gardens and into a stew: "Towards the Cutraine then you must be gon, / The garden alleyes paled on either side, / Ift be too narrow, walking there you slide. / Into a house among a bawdy crew, / Of damned whores; I theres your whole delight."[76] For West, the pale and narrow garden alleys of Shoreditch lead to different kinds of entertainments, fostering associations between urban gardens, theatrical performance, and vice. Associations between Bankside's gardens and stews worked in similar ways; the pleasures of the Little Rose were undoubtedly linked to the scents emanating from its two adjacent gardens.

The Paris Garden—one of the city's two public amphitheaters used for bull- and bearbaiting—offered its own powerful scent, defining the area's association with theatrical (and odiferous) gardens. In Jonson's *Epicoene*, Truewit threatens to send his wife "ouer to the banke-side" after she criticizes his rough manners: "Must my house, or my roofe, be polluted with the sent of beares, and buls, when it is perfum'd for great ladies?"[77] Perfume, associated with aristocratic interiors, is juxtaposed with the "pollution" of animal scents associated with public entertainments. The joke, of course, is that such juxtaposition was hardly contradictory. By the 1590s, most early modern perfumes were animal based and worn on leather: the scent of civet, musk, and cats rivaled those of bears and bulls. The amalgamated scent of roses, stews, bears, bulls, and players that defined Bankside within contemporary sources (many of them, ironically, plays staged in Bankside) performed an olfactory logic that manifested itself in 1613 with the building of the Hope Theatre, the first theater in London designed to house both baiting and drama. With its covered roof and moveable stage, The Hope was a unique venue, with its own particular smell, which helped Jonson dramatize other unsavory zones of the city, namely Smithfield market in *Bartholomew Fair*.[78]

The smellscape of Bankside included odors emanating from the diverse entertainments found in its various gardens. A character in Thomas Dekker's *Satiromastix* (1601) describes another's breath "as sweet as the Rose that grows by the Beare-garden," a compliment that draws on localized, spatial knowledge while also emphasizing the proximity of both roses and bears. The scent of one heightens the presence of the other. Thus, the area's rose gardens, stews, and baiting arenas all define the "place" of London's Rose Theatre. Just as rose perfume enabled Henry VIII's kingly body to extend into space, the scent of roses in Bankside helped locate theatrical performances within a particular environment, connecting the social world of the stage with those of early modern London audiences.

Dekker's *Shoemaker's Holiday* (1599) is one such play that accomplishes this feat; Shakespeare's *Romeo and Juliet* is another. Both plays focus on young love forbidden by their respective social worlds, and both stage erotic identity through puns on distillation and perfume. In Dekker's play, Rose, the daughter of the Lord Mayor and perhaps the first English character with such a name, falls in love with Rowland Lacy, an English courtier. Though it is set in the fifteenth century, Dekker's play stages sixteenth-century commodity culture—a stunning array of silks, taffetas, and cloth circulate throughout the play alongside a number of perfume references. Banished to the country, Rose sits alone and makes a garland of flowers for her love, Lacy. Though she is "imprisoned" in her father's house, she imagines herself on a "flow'ry" bank of "pinks," "roses," "violets," "gilliflowers," and "marigolds" (2.1.5).[79] Later, when she elopes with her lover, Rose's identity is confused with another woman's, inspiring Fisk, a journeyman shoemaker, to anachronistically joke about the mistaken identity of "Mistress Damask Rose," a linguistic pun that gestures toward an evolving notion of bodily presence dependent on olfaction and perfumes. Rose's name literally becomes the damask rose. She is also referred to in the play as the "little" Rose, a knowing wink to the spatial environs of Bankside and to the nearby brothel. Finally, when given a three-penny piece by another character, Fisk remarks that he "smells the rose," a phrase that has been interpreted by editors to refer both to the materiality of money—Fisk smells both the Tudor three-penny piece discussed above (with the symbol of the rose) and where he will soon spend it—in a nearby tavern and brothel (the little Rose). Fisk's series of puns on the many roses in the play—of Rose's name, of currency marked with its symbol, and of Bankside stews and theaters—connect the play's fifteenth-century "city" with early modern London. Dekker's Mistress Damask Rose thus shows how, by the end of the sixteenth century, the damask rose's floral scent was an integral part of staging embodied desire within the social world of the stage and in early modern London.

Like Dekker's Rose, Shakespeare's Juliet also demonstrates the scope of the appeal of perfumed roses, famously linking their sweet scent to *Romeo and Juliet*'s broader examination of love and its transformative role in identity. As critics note, the play's staging of eroticism (and the confusion of identity that it inspires) was, in large part, inspired by Hall's account of Henry VIII's and Anne Boleyn's erotic revelry in *The Assault on the Castle of Virtue*.[80] First published in 1597, the play examines how theatricality intersects with the social world of Verona through staging its violent streets, its domestic interiors, and, importantly, its gardens. Catherine Belsey argues that *Romeo and Juliet* explores the embodied effects of individual desire on society and culture as a whole by examining the relationship between bodies and culture. She begins with a question about spatial orientation: "Is the human body inside or outside culture?"[81] Or, put another way, is love a "matter" of the body or an abstraction of the mind? Whereas Belsey uses this framework to argue that the play "imagines a metaphysical body that cannot be named," I maintain that such an imagined metaphysical body can be sensed and that the scents of Bankside's roses provide useful ways of approaching the play's orientations of desire.[82]

Examining Juliet's articulation of the (anticipated) pleasures of sex, Belsey argues that metaphors of sensation are an important part of the play's struggle between passion and identity. Juliet, alone in her house, waiting for Romeo, muses on the physicality of sex, addressing "[n]ight" to "come" and "learn" her "how to lose a winning match" (3.2.12).[83] If the play stages love at first sight, then it equally stages desire at first touch. As Belsey argues, Juliet addresses "[n]ight" in order to foreground sensory pleasures: "It is as if Juliet imagines the presence of the desiring bodies as pure sensation, sightless, speechless organisms, flesh on flesh, independent of the signifier. A rose by any other name, she had earlier supposed, would smell as sweet: the same gases, emanating from the same petals, striking the same nostrils, its physical being separable from the word that names it. The name, the signifier, and the symbolic order in its entirety are to be relegated to a secondary position, the place of the merely expressive and instrumental . . . But these isolated, unnamed bodies (and roses) are only imaginary."[84] Romeo and the night are conflated in an erotic celebration of the paradoxical, exquisite pleasures of consummated passion: "Come night, come Romeo: come, thou day in night" (3.2.17). Yet, as Belsey argues, such a pure sensation cannot be staged, only imagined and described: the isolated, unnamed bodies, like the perfumed roses, can only even be "imaginary."

Although Belsey's brilliant reading demonstrates how the play stages the impossibility of metaphysical, idealized love, it does not consider how desire and sensation unfold within spatial coordinates—both in the imagined world of Verona and on the stage in London. Why must these roses—and their scent—be imaginary? What happens to such articulations of desire when one factors in the smellscapes of Verona and Bankside, including those odors emanating from nearby rose gardens? And how does the smell of the rose—as a metaphor for identity and anatomy—work given the play's fascination with distilled essences?

Both in London and in Verona, gardens were imagined as somewhere in between public and private space, an apt setting for theatrical (and olfactory) self-fashioning. John Norden's 1593 map of London depicts the realm between the Paris Garden and the Rose Theatre as blank space, dotted with trees. Bankside, in contrast to the crowded, walled city to the north, is represented on the map as a primarily agricultural zone, with a row of stews to the north and a thoroughfare from the south that leads to both the Rose and the Paris Garden. The rest of the space is small lots with a few dwellings. Likewise, in Verona, vegetative space borders the Capulet home. As Romeo walks home from the Capulet masque after his clandestine exchange with Juliet in the garden/balcony scene, he encounters Friar Laurence scavenging to fill up his "osier cage" with "baleful weeds and precious-juicèd flowers" (2.2.7–8). Staged in a threshold zone—somewhere between Verona's public streets and its private households—this scene foregrounds nature as part of urban space.

The editors of *The Norton Shakespeare*, for example, note that the scene takes place within the walled space of the city but not within the walled space of the Capulet household or its gardens, which Romeo has just left. In this liminal zone, the Friar muses not on the effects of desire and sensation but rather on the

John Norden's Map of London (1593). Depicts the realm between the Paris Garden and the Rose Theatre as blank space, dotted with trees. © The British Library Board, Maps.Crace.1.21. All rights reserved 06/08/2010

effects of herbal and pharmacological essences: "O mickle is the powerful grace that lies / In plants, herbs, stones and their true qualities" (2.2.15–16). Noting that vice can be a virtue, at least in herbaceous form, the Friar makes an argument for distillation's transformative powers, particularly in making medicinal perfumes. It is at this moment that Romeo enters the scene, and the Friar's speech about "weak flowers" gains additional resonances: "Within the infant rind of this weak flower / Poison hath its residence, and medicine power, / For this being smelt, with that part cheers each part; / Being tasted, slays all senses with the heart" (2.2.23–26). Musing that both "grace and rude will" reside in the weak flower, the Friar notes that sensory encounters matter: the flower, if smelled, is a virtue. If tasted, it "slays" all sensation.

Romeo's entrance at this moment in the speech links his embodied passion for Juliet to the embodied reactions to the distilled flower described by Friar Laurence.[85] Indeed, the men of Verona are repeatedly compared to flowers, raising interesting questions about whether erotic attraction is based on an intrinsic, subjective perception of beauty or on some extrinsic, outward manifestation of it. Juliet's mother describes Paris as a rare floral specimen: "Verona's summer hath not such a flower," a point echoed by the Nurse: "Nay, he's a flower, in faith, a very flower" (1.3.78–79). Compared to Paris, Romeo "is not the flower of courtesy," at least in the Nurse's estimation (2.4.42). Chiding Romeo for giving him the slip at the Capulet masque, even Mercutio, too, puns on the relationship between courtesy and flowers, describing himself as a "courteous" pink carnation. Romeo then boasts that his "pump" is well "flowered," linking both the carnation's color to female genitalia and the tactility of "pinking" or perforating something to the physical act of sex. Juliet uses a metaphor of a flower to describe her amorous encounter with Romeo in the garden, though her phrase uses floral scents to imbue a futurity to their love: "This bud of love by summer's ripening breath / May prove a beauteous flower when next we meet" (2.1.163–65).

Such metaphors most likely had olfactory referents: stage directions call for spices, herbs, flowers, and sweet-waters, especially in the play's final act. References to stage properties at the Rose suggest that there was an abundance of materials to stage trees, moss, and banks, though the nearby gardens could just as easily provide floral specimens. The olfactory presence of roses on stage would heighten the tragedy of Juliet's death. Like the fragrant rose in Sonnet 54, Juliet's untimely death produces its own perfume on stage, which her family interprets as the scent of her chastity; her father, discovering her death, calls her a "flower . . . deflowered" by death. Paris, mourning Juliet at her grave, addresses her by saying "sweet flower, with flowers thy bridal bed I strew." Of course, Juliet is not dead, nor is she a virgin; floral perfume is thus linked with the young lovers' erotic transformation and theatrical ruse.

The play also stages a variety of distilled essences. The Nurse calls for aqua vitae after reporting Tybalt's death. The soporific agent that Friar Laurence gives to Juliet is described as a "distilling liquor," and Juliet refers to it as a "vial." Romeo seeks out the apothecary, who, among "cakes of roses" and other "musty" herbs and seeds, "cull[s] simples" (5.1.46–47, 40). Finally, Paris, sprinkling distilled sweet-waters on Juliet's grave, offers her his own tears, "distilled" by moans (5.3.15). Thus, Friar Laurence's meditation on the distillation of floral matter—and its bodily effects—resonates across the play, demonstrating that sensory orientations matter within the plot of the play. To borrow Juliet's metaphor from the garden scene, part of the problem is that the young lovers' passion is not allowed to bloom on its own. Just before their nuptials, Romeo imagines his future happiness through spatial sensation: "if the measure of thy joy / Be heaped like mine, and that thy skill be more / to blazon it, then sweeten with thy breath / This neighbor air, and let rich music's tongue / Unfold the imagined happiness." (2.4.24–28).

Yet the young lovers' happiness does not unfold as they imagine; rather, like one of the Friar's "baleful weeds or precious-juicèd flowers," it is consumed too early and "being tasted, slay[s] all senses with the heart." Staging a conflict between individual, erotic, embodied passion and social identity, *Romeo and Juliet* queries what is symbolically represented by a name. In the play's balcony scene, in which Romeo and Juliet seek to reconcile their embodied desire with a sense of self, Juliet, under cover of night on the balcony above the Capulet's walled gardens, famously muses: "What's in a name? That which we call a rose / By any other word would smell as sweet" (2.1.82–87). Juliet's faith in the transhistorical, transcendent sweet scent of a rose demonstrates that, by the early 1590s, roses were valued for (and perhaps synonymous with) their scent. The rose that Juliet so blithely describes was undoubtedly a damask rose, its sweet perfumed scent allowing her to ponder what was essential about identity. So, too, does Romeo: later in the play, desperate to imagine a way for his doomed love to flourish, he pitifully begs the Friar to tell him "in what vile part of this anatomy / doth my name lodge? Tell me, that I may sack / The hateful mansion" (3.3.106–7). But the play refutes such a physical approach, staging instead an olfactory and pharmacological investigation of desire. As Friar Laurence's role in the plot emphasizes, distillation—not anatomy—provides a resolution, albeit one that is all consuming, that "being tasted, slay[s] all senses with the heart." The social world of Verona is transformed, but only through the sacrifice of the two young lovers: Juliet's sweet perfumes materialized on stage through funerary flowers and distilled perfumes.

Both Henry VIII's dramatic performance of self and the kinds of theatrical performances staged in Bankside emphasized a phenomenological experience of embodiment that was increasingly part of Renaissance self-fashioning. That both utilized rose perfumes to do so raises important questions about the role of the senses in materializing the body and its boundaries. Rose perfume was an important part of changing conceptions of the self because it extended one's bodily presence into space and provided a way to conceptualize the body's effect on its environment. In essence, rose perfumes enabled a phenomenological experience of English embodiment that ironically depended on mastering foreign ingredients and technologies. Whereas Henry VIII utilized rose attar to embody a powerful kingship that depended less on his divine presence and more on his physical form, stage players depended on the scent of rose perfume to locate their performances within social worlds that resonated on stage and in the community of Bankside. Both cases suggest that olfaction captured a profound moment of contact—with a king or with a different social world. Rose perfumes thus connected an emerging English national identity with subjective, phenomenological experiences of sensation. As the next two chapters demonstrate, however, this olfactory extension of the body's boundaries was not without problems, since such contact was never one-sided in airborne environments. The smell of sassafras marked one environmental peril, underscoring the paradox of English olfaction in non-English realms, specifically in Anglo-Indian contact zones. Rosemary was another, its sweet scent a sickeningly powerful reminder of the persistent presence of the plague in London's crowded spaces.

CHAPTER THREE

Discovering Sassafras

Sassafras, Noses, New World Environments

Sir Humphrey Gilbert, knight, merchant adventurer, veteran of English wars in Ireland, and an important early proponent of English colonization of the Americas, argued strenuously that English success in the new world required blind faith in the endeavor. In his widely read *Discourse of a New Passage to Cataia*, which circulated in manuscript form for almost a decade before it was published without his permission in 1576 by George Gascoigne (and later republished in Richard Hakluyt's *Principal Navigations* in 1589), Gilbert argued that the bold would be separated from the brute by their ability to transcend their senses: "The diversity between brute beasts and men, or between the wise and the simple, is, that the one judgeth by sense only, and gathereth no surety of anything that he hath not seen, felt, heard, tasted, or smelled: and the other not so only, but also findeth the certainty of things, by reason, before they happen to be tried."[1]

In the transcendence of the senses lay Gilbert's prescient "certainty" of English colonialism. Although English endeavors in the new world hitherto had been unsuccessful, rendering only a profound sensory absence—things *not* seen, felt, heard, tasted or smelled—Gilbert was sure that undetected riches would reward those who persevered despite such a lack of evidence. As the byline of the *Discourse* implies, no goal was impossible: "Sir Humphrey Gilbert, Knight, *Quid Non?*"[2]

Perhaps recognizing the temerity of such a motto and the need for broader support of English colonization, Gascoigne, in his introduction to the *Discourse*, supplemented Gilbert's stark insistence on reason with a plethora of sensory embellishments. Likening himself to a "huswife" who "is no lesse curious to decke her bees hive, to rub and perfume it with sweet herbes, to cover and defend it from raine with clay and boordes, and to place it in the warme Sunshine safe from the Northerly blastes," Gascoigne thought it "good (Allegorically) to write in behalfe of the right and worshipful and [his] very friend S. Humfrey Gilbert. . . . As a good Gardener doth cover his tender herbes in winter, and cherishe them also in summer: so have I thought my selfe bounded somewhat to . . . answere vnto the obiections that might bee made."[3] Redefining Gilbert's sensory dearth through allegorical perfumes, Gascoigne argues that, unlike other texts which describe things "already known," Gilbert's "doth tend to a very profitable and commendable practice of a thinge to be discouered."[4] Gilbert's emphasis on the absence of sensory data, and Gascoigne's allegorical buzzing and perfuming in the wake of such a claim, shows the ways early English colonial endeavors hinged on desire for sensible proof of things yet "to be discouered."

Gilbert was poised for success. His goal was to claim Newfoundland, an area well known to English (and French, Portuguese, Spanish, and German) fishermen, for England.[5] As Gilbert prepared for his voyage, he petitioned alchemists, mathematicians, and natural philosophers for advice, reading extensively in the classical and contemporary accounts of the flora, fauna, navigational routes, and peoples in the new world.[6] In 1578, he secured a charter from the queen; he invested his entire fortune, along with his wife's, in the endeavor. In 1583, Gilbert set sail for Newfoundland with 260 men, including apothecarists, carpenters, masons, physicians, refiners, a band of musicians, and sundry "petty haberdasherie wares to barter with these simple people."[7]

Despite such extensive preparations, Gilbert's trip was brief. Early conflicts with fishing fleets off the coast of St. John's Bay, combined with short supplies, hastened his return to England.[8] And though one refiner purportedly discovered silver mines, all samples and notes were lost when the ship he was on sank off the cost of the Sable Islands. Gilbert headed back to England empty-handed, determined that the reports alone could convince the queen to invest heavily in future endeavors.[9] On the return voyage, they encountered rough seas near the Azores. His ship, the *Squirrel,* sank, killing all onboard. Gilbert reportedly remained resolute in his conviction of England's divine manifest destiny until the very end. He was last seen on deck, holding a book in his hand, quoting Thomas More: "we are as neere to heaven by sea as by land."[10]

Gilbert's journey was a spectacular failure, but not in a way he or Gascoigne anticipated.[11] Gilbert had prepared for a dearth of sensory data, but he encountered its opposite—an overwhelmingly harsh and inhospitable sensory environment. One participant recorded an "impassable" wilderness, cluttered with undergrowth and fallen trees, with unbearable heat in summer, insurmountable "heapes and mountaynes" of ice in winter, and "ayres" riddled with mists and constant rain.[12] A later account corroborates this point: Newfoundland's airs were thick, corrupted with the vapors of half-rotten trees; the roar of the sea so great that it could drown out the sound of canons; and its natives were cunning and fierce, described as "wolves."[13] Worse still, it seemed devoid of merchandisable commodities. Southward, ever southward, lay English hopes in the new world.[14]

On Gilbert's death, his half-brother, Sir Walter Raleigh, inherited the charter.[15] Learning from Gilbert's mistakes, Raleigh abandoned Newfoundland for more temperate land.[16] In 1584, Raleigh sent two servants, masters Philip Amadas and Arthur Barlowe, on a four-month reconnaissance mission of new coasts, led by veteran Captain Simon Ferdinando, a pilot under Gilbert's command during his first attempted voyage to Newfoundland in the 1570s. Perhaps wary of Gilbert's northern route, Ferdinando risked encountering the Spanish fleet and headed farther south, stopping briefly in a secluded harbor off a large island in Spanish-held West Indies (probably Puerto Rico), before heading north of Spanish territories in Florida toward the Carolina coast.

Such risk seemed to offer immediate reward. In June 1584, Amadas and Barlowe detected a scent wafting out to sea; it was a smell "so sweet and so strong . . . as if we had bene in the midst of some delicate garden abounding with all kinds of odoriferous flowers, by which we were assured, that land could not be farre distant."[17] Two days later, they reached land and discovered the origins of that smell: cedar, pine, Lentisk (or Mastic, which they erroneously identified as cinnamon), and sassafras, which made a "most plentifull, sweete, fruitfull and wholesome [soile] of all the worlde: there are above fourteene severall sweete smelling timber trees, and the most part of their underwoods are Bayes and such like: they have those Okes that we have, but farre greater and better."[18] The voyage was reported as a great success: in their accounts, Barlowe and Amadas describe the Carolina coast as a veritable paradise, peopled with peaceful natives, temperate climes, and abundant resources. More importantly, they returned with proof.[19] Based on their reports, Raleigh readied for the first English colony in the new world in Roanoke.[20]

Sensory Worlds

It is tempting to read these two accounts of the new world against one another: whereas Gilbert's is a tale of failure that requires Gascoigne's allegorical perfume, Barlowe and Amadas's account reports overwhelming new world material sensations, including a most fragrant terra firma. Gilbert hopes for sensations yet to be discovered; Amadas and Barlowe describe new world delights experienced firsthand. Gilbert's is a tale of northern failure; Barlowe and Amadas's opens Raleigh's southern plantation saga. However, both accounts approach the new world as sensory zones, describing overwhelming, new environments through multisensorial data. Though Gilbert's impact at Newfoundland was ephemeral, it greatly affected how sixteenth-century English men and women understood America and its landscape, whether imagined as allegorically perfumed or experienced firsthand as cold, harsh, and vast. This was especially true for the explorers moving between the contact zones of Newfoundland, Roanoke, and Jamestown, who learned that successful colonization in the new world would require mastering such environs.

Describing these colonial failures as early lessons in "sensory warfare," historian Peter Hoffer argues that such encounters shaped English approaches to the colony in Virginia.[21] Faced with "oceans" of forests and the "roar" of the sea, explorers increasingly relied on sense of smell, which offered a proximate way through unfamiliar terrain. As the English searched for sensible, and merchandisable, matter in these realms, they learned that their failure and success depended on mastering new strategies of "discovery," including olfaction. One smell in particular symbolized for the English the potential for success in the new world—the smell of sassafras.

Sassafras may seem a rather surprising commodity for the English to have based their early endeavors in the new world, especially since its legacy has been eclipsed by tobacco. But for a brief period of time, before successful cultivation of tobacco at Jamestown in 1622, sassafras seemed to offer the English a way to excavate botanical riches from new world ecologies, just as the Spanish had a century before. Though a host of perfume ingredients fueled English commercial interest in the new world, sassafras in particular gripped English imagination. Between 1602 and 1607, six separate voyages went in search of it in the region south of St. Lawrence and north of Roanoke in order to meet commercial demand.[22] Growing along the eastern shore of North America, in Florida, Virginia, and New England, it galvanized European markets in New Spain, New England, and New France as a potential cure for those suffering from the pox or syphilis.[23] Even as smallpox decimated native populations in the new world, the great pox, believed to be an exogamous disease with origins in the new world, raged in Europe. Galenists believed that sassafras, or "ague-tree," could relieve ailments of the liver or spleen and, like tobacco and guaiacum, cure syphilis.

Yet the aromatic qualities of sassafras associated it with the "contagion" of luxury perfumes and the sensuality they connoted. In Ben Jonson's *Cynthia's Revels*, Amorphous and his perfumer (Signior "Fig") discuss a price for Spanish scented leather gloves. When Mercury, disguised as a page for the courtier Amorphous, inquires of their value, the perfumer names no set price: "Give me what you will, sir: the *Signior* payes me two crownes a pair; you shall giue me your loue, sir" (5.4.400).[24] Mercury, outraged perhaps at the price and at the suggestion, exclaims, "My loue? with a pox to you, goodman *sasafras*" (5.4.399) Love leads to the pox. The perfumer, formally an aphrodisiacal "fig," becomes "Goodman Sassafras." The joke here is multifold. Mercury was a rival cure for syphilis in the period, often prescribed by Paracelsians; Mercury allegorically haggles with sassafras, capturing the widespread cultural link between the pox and its cures. The perfumer, however, does not reject his status as "Goodman Sassafras"; he merely reconfigures his approach to making a sale. Shifting from gloves to pomanders, he playfully responds, "I come, sir. There's an excellent *diapasme* in a chaine too, if you like it." The perfumer sells both aromatic aphrodisiacs and cures, namely sassafras, for the diseases they inspire, namely syphilis.[25]

Such a joke may seem surprising to modern readers. Within modern sensory configurations, sassafras is most commonly identified as a taste rather than a smell, let alone a perfume ingredient.[26] As Jonson's play documents, however, sassafras was valued for its aromatic scent in early modern Europe. Like rosewater, sassafras was believed to have powerful medicinal, religious, and ornamental uses. Its bark was distilled into an essential oil that smelled like fennel; its leaves were also distilled, producing a citrusy essential oil; and medicinal teas made from crushed leaves offered their own scent, similar to bergamot.

During the last quarter of the sixteenth century, sassafras was a panacea; in some accounts, it was even reported to protect the wearer from aging. By the end of the first quarter of the seventeenth century, it was made into infant cribs and aromatic boxes to house English bibles, due to its purported ability to ward off evil spirits. Ships with large amounts of sassafras in their hulls were believed to be unsinkable. The scent of sassafras, described in early-sixteenth-century sources as rivaling the scent of the Ceylon cinnamon tree, was thus one of hope, particularly for consumers in England. But, as I argue, for English explorers, it was also a scent of hunger, failure, and loss, gesturing to the complicated ways in which the scent of the new world was experienced both in the contact zone and in Europe. Its varied smells underscore the seemingly limitless medicinal and ornamental uses for sassafras in Europe and the inherent difficulty in finding it in the new world.

What is most striking about the role of sassafras in shaping European encounters in the new world is that it was so ubiquitous. Christopher Columbus, in some myths, was drawn to terra firma by its scent. Samuel de Champlain recounts its succor and delightful smell in his explorations of Maine. Francisco Hernández describes it in his travels through Mexico. Nicolás Monardes publicizes its role as a panacea in his *Joyful Newes*. Jacques le Moyne de Morgues envisions it encircling Jean Ribault's column in eastern Florida. Thomas Harriot proclaims it one of three important commodities native to Roanoke. John White's watercolors depict natives literally covered in it. Ralph Lane claimed to survive famine by drinking a tea made with it. Martin Pring, under the guidance of a collaboration of Bristol merchants, led a trip to Newfoundland solely to collect it. Walter Raleigh was convicted of treason when his covert continental correspondence about offloading a stolen cargo of it was misinterpreted as part of the Wye rebellion. It is one of eleven commodities specifically mentioned in the Virginia Company's charter for Jamestown, and it was one of the first exports Captain John Smith sent back to England. By 1620, members of the Virginia Company complained that planters were too focused on cultivating tobacco and sassafras for private profit, at the expense of company lands.

Despite being so pervasive in early accounts, sassafras has faded from most scholarly accounts of the new world and Anglo-Indian contact. Of all the scent ingredients examined in this book, sassafras is perhaps the most ephemeral. Though a lucrative commodity, the difficulty of harvesting it made it an unreliable one on which to base permanent English colonies. At times, it was rare and valuable in English markets; at other times, it was plentiful and cheap. As such, it often failed to bring in the hoped-for profits, particularly as a perfume ingredient. It would not be until the end of the seventeenth century, with the introduction of jasmine, that a distinctively new world perfume would emerge in Europe for mass consumption (see chapter 6). Rather, sassafras, as a perfume ingredient, documents only evaporated hope for early English voyages to the new world and would soon be replaced by the smoke of tobacco. As such, its ephemeral history gestures toward the lost sensory worlds of Anglo-Indian contact and the ways olfaction functioned within them.

English sassafras consumption may seem only a curious historical footnote until one considers its large role in shaping Anglo-Indian contact. In order to understand how the new world, repeatedly described in sixteenth-century sources as an overwhelmingly fragrant land—a veritable Eden—became, in seventeenth-century ones, a savage and hostile "wasteland" requiring deforestation in order to establish English tobacco plantations, we must grapple with the sensory worlds of the contact zone and the challenges they posed to European explorers.[27] In the nascent years of English colonialism, before proofs of success were visible, sassafras represented a paradox of sensation for Europeans in the new contact zones. When one switches from technologies of sight to technologies of olfaction, the phenomenology of discovery becomes proximate and dangerous.

Sassafras was not easy to find, nor was it easy to harvest. Early reports sent back to Europe described only the plant's roots, making it difficult to identify living specimens by sight alone. Most adventurers had only vague descriptions of the sassafras root to guide them, with almost no description of the tree or flower.[28] Nicolás Monardes's depiction of it was reprinted widely, but sassafras varies greatly along the eastern seaboard. Small in the north, it could grow to almost 80 feet in the south, with a trunk six feet wide.[29] The leaves also change shape: old boughs have boat-shaped, smaller leaves while new ones are large and mitten-shaped. As explorers soon discovered, one of the only ways to identify it was through the scent of its leaves, a method that sounds simple enough until one factors in the dense "oceans" of forests (and their overwhelming collective scent) described in travel narratives. The experience of searching for sassafras, along with other botanical matter, was often frustrating. Francis Higginson wrote from Boston in 1629, "This country abounds naturally with store of roots of great variety . . . and other sweet herbs delightful to smell, whose name we know not."[30] As such accounts document, harvesting sassafras required a deep sensory engagement with a foreign ecology. English explorers seeking lucrative sassafras (of which there were many) had only their noses, or native informants' noses, as guides.[31] Sassafras challenged English visual mapping of landscapes. Success of this crop would require more than visual proofs; it would require olfactory proof, rendering English bodies vulnerable to the new environs of North America.

The first European reference to sassafras occurs in the second part of Nicolás Monardes's influential three-volume catalog of new world drugs, *Historia medicinal de las cosas que se traen de nuestras Indias Occidentales* (1569). Republished in an expanded version in 1574, and translated into Latin and English shortly thereafter, Monardes's text was influential and worked to connect new world ingredients to established European medical practices.[32] Arguing that the new world's vast riches were pharmacological rather than metallic, Monardes cited four drugs in particular: tobacco, guaiacum, sarsaparilla, and sassafras. Whereas scholars have examined the impact of tobacco and guaiacum on European medicine and consumerism in great detail, the impact of sarsaparilla and sassafras remain understudied.[33]

The value of both stemmed from their use as medicinal cures.[34] In his English translation of Monardes's book, John Frampton emphasized that the treatise on new world materia medica could also function as a kind of field guide for merchant adventurers. Renaming it *Joyfull Newes out of the Newe Founde Worlde*, and, more importantly, adding "portrature" of the "said hearbes, very aptly described," Frampton asserted in his title that Monardes's treatise "might bring in tyme rare profite, to my Countrie folks of England."[35] Frampton, an English merchant in Seville, turned translator on his return to Bristol after he was captured, imprisoned, and had his goods confiscated by the Inquisition.[36] He was notoriously reticent to discuss his time in Spain; his translations, some have suggested, represent an attempt at revenge, disseminating Spanish knowledge about the new world to English merchant adventurers like him.[37] Frampton states in his introduction that medicines Monardes includes, like tobacco and sassafras, are already merchandizable: "[They are] now by Marchauntse and others, brought out of the West Indias into Spaine, and from Spain hether into England, and that the excellencie of these Hearbes, trees, oyles, plantes and stones, &c. hath been knowen to bee so precious a remedie for all maner of deseases, and hurtes, that maie happe unto Man, Woman, or child."[38] English profit lay in mastering Spanish trade networks and experience with new world materia medica, like sassafras.

After writing seven pages on the wonders of tobacco and two on the power of nicotine, Monardes begins his eleven-page description "of the Tree that is brought from the Florida, whiche is called Sassafras" by asserting its great "vertues" and "excellencies." These he corroborates through firsthand reports, citing sailors in Florida, doctors in Havana, and opportunists in Michoacán, all of whom confirm his assessment of the virtues of this "woodd and roote."[39] He also relies on his own experiential knowledge of the root, suggesting that he had specimens in his famous Seville garden: "It maie bee three yeres paste, that I had knowledge of this Tree, and a Frenche manne which had been in those partes, shewed me a peece of it, and tolde me merueiles of his vertues."[40] At first, Monardes remained skeptical of the French man's claims, since "of these thynges of plantes, and hearbes, whiche is brought from other places, thei saie muche, and knoweth little," but he vouches for the efficacy of sassafras: "now I dooe finde to bee true, and haue seen by experience."[41] By emphasizing the visual nature of sassafras— "seeing" by experience—Monardes helps skeptical readers also "see" the "truth" of its efficacy through the circulation of print.

Such claims of experiential knowledge of new world botany bridged the sensory worlds of the contact zone and the metropole. Like rose attar, which extended the body's influence on an environment through aromatic self-presentation, the smell of sassafras extended the influence of the environment on the body. When the French Huguenots, led by Jean Ribault and René Goulaine de Laudonnière, were massacred near the St. Johns River by the Spanish led by Pedro Menéndez de Avilés in 1565, only a few survived. Those who did warned the Spanish of a terrible disease—a tertian fever—with an efficacious, local cure: sassafras, about which the Timucua of Florida (Outina) had informed them. Monardes writes,

Nicolas Monardes, John Frampton *Joyfull Newes Out of the Newe Founde Worlde*
(London, 1577). Widely reproduced in seventeenth-century texts, including maps of
Virginia, this image helped guide English explorers in their search for sassafras.
By permission of the Folger Shakespeare Library

"[A]fter the Frenche menne were destroyed, our Spaniardes did beginner to waxe
sicke, as the Frenche menne had dooen, and some which did remaine of them,
did shewe it to our Spaniardes, and how thei had cured them selues with the
water of this merueilous Tree."[42] Sassafras returned to Europe with the conquer-
ing Spaniards: "[I]n this tyme it was, when the Capitaine generall Peter Mellen-
dis came from the Florida, and brought with hym in common this wooded of the
Sassafras."[43]

Buried in Monardes's description of sassafras is a glimpse of the conflict and exchange between the Timucua, French, and Spanish in Florida—and, in Frampton's translation, between the English and Spanish. Both a commodity to be harvested and a panacea that underscored the dangers and diseases latent in new environments, sassafras was a complicated symbol, useful in both new and old worlds. From a modern perspective, it may seem emblematic of European hysteria and greed, along with outmoded medical practices, but to early English explorers, it represented both the risk and reward of the unfamiliar terrains, weather, flora, fauna, and peoples of the new world.

One of the most famous depictions of cultural exchanges between the Timucua, French, and Spanish is Jacques Le Moyne de Morgue's depiction of Athore, a Timucuan chief, showing Laudonnière Ribault's column. The role of sassafras in this iconic image has not been examined closely. Le Moyne accompanied Laudonnière to Florida, commissioned by King Charles IX of France to chart and map the country and document its people. The majority of his drawings, however, were lost in the Spanish attack on Fort Caroline; Le Moyne was one of only two survivors who escaped the fort with their lives and regrouped with the French fleet out at sea.[44] He eventually ended up in London, where he re-created his sketches from memory, circulating them among English adventurers, translators, and printers like Raleigh, Harriot, White, and Theodor de Bry. The *Narrative of Jacques Le Moyne de Morgues* was published in de Bry's *America, Part II* in 1591 along with illustrations based on re-created sketches, out of regard for "intelligent people," so that "the story should not just be told but should also seem to be enacted vividly before their eyes."[45] Such sensory evidence documents an attempt to render the overwhelming sensations of the contact zone tangible and manageable, in particular the invisible yet powerful commodities like perfume ingredients.

The vivid spectacles depicted include de Bry's engraving of Timucuan "worship" of Ribault's column. Like Gilbert's cross in Newfoundland, Ribault's column served as a visual reminder of colonial failure rather than success. It was planted ten years before, as Jean Ribault struggled, and failed, to establish a colony.[46] In Le Moyne's narrative, Laudonnière returns to Ribault's abandoned Florida fort, discovering that the natives have reclaimed the stone, "worshipping it as if it were an idol."[47] Significantly, such recovery could also be understood as perfuming. Le Moyne describes various offerings of "fruits of the district and roots that were either good to eat or medicinally useful, dishes full of fragrant oils, and bows and arrows; it was also encircled from top to bottom with garlands of all kinds of flowers, and with branches of their most highly

Jacques Le Moyne de Morges, *Laudonnierus et rex athore ante columnam a praefecto prima navigatione locatam quamque venerantur floridenses* (1564). A sassafras garland adorns Ribault's column, just underneath the French arms, emphasizing its value in Florida and in Europe. Print Collection, Miriam and Ira D. Wallach Division of Art, Prints, and Photographs, The New York Public Library, Astor, Lenox, and Tilden Foundations

prized trees."[48] Anthropologists are wary of reading too much into this representation of the Timuca, noting that Le Moyne and de Bry represent the natives with blond hair and white skin.[49] The offerings in the foreground, like the whitened natives, are equally out of place, a cornucopia of European baskets filled with European fruits. In some ways, de Bry's picture offers "intelligent" readers only a compelling European fantasy of dominance over simple people and fertile lands, filled with recognizable commodities. But scholars corroborate that the metal pendants on Arthore's belt were native to eastern Florida. And the middle garlands of the natives' "most prized branches" resemble sweet bay and sassafras.

Though it is impossible to identify these specimens with certainty, the resemblance is striking. The image of the natives "perfuming" Ribault's column offers visible proof of sassafras's role in both new world spaces and European imagination about them. It works both as anthropological evidence for a phenomenology of the contact zone and as evidence of cultural refractions of English endeavors in the new world; after all, de Bry's engraving appeared in print in 1591, after White's trip and return to Roanoke. Le Moyne retired to England, where his new world expertise connected him to Raleigh's influential circle, including Harriot, Hakluyt, White, and Sir Philip Sidney. With Mary Sidney's patronage, he published his *La clef des champs*, a collection of European plants. In the *Narrative*, sassafras serves as a warning; the French survivors momentarily align themselves with the Spanish in order to warn them of diseases latent in the environment, hinting at the importance of native knowledge about its cure. Embedded within Le Moyne's image is a phenomenological trace of the contact zone, and the ways the smell of sassafras remained a reminder of the perils of exploration. Yet when this narrative is depicted in de Bry's *America*, and "brought to life" for European eyes, sassafras is no longer a smell but an image. For most European readers, it no longer offered a warning but rather symbolized the riches of new landscapes and the pliancy of their inhabitants. It represented hope and financial gain, not dangers of breathing in foreign environments. Such sensory shifts would shape the history of Anglo-Indian contact at Roanoke and Jamestown.

A View of Virginia's Invisible Sensory Realms: White's Sketches of Sassafras

The English were slow to adapt to the frighteningly new sensory realms of the Atlantic coastline. Like Newfoundland's inhospitable, frozen landscape, Roanoke's dark forests, thick fogs, noisy animals, and sweet airs presented a difficult environment to master and control, particularly using visual technology. Later, near Jamestown and in Martha's Vineyard, the English would learn from these early mistakes and shift tactics, burning large portions of forests in order to assert greater visual control over the areas surrounding their camps. Though the goal was obviously to clear the land, the resulting smoke from such fires also functioned as sensory warfare—a visual and olfactory assault—upending the sensory hierarchy of the contact zone that very much favored the natives. Natives would have been able to smell such smoke miles away. Its stench thus extended the colonists' presence far beyond their armed forts, sending a clear and menacing signal of impending violence to natives hidden nearby in the forest.

Such tactics reflect a profound adjustment in English approaches to the new world, shifting from discovery and exploration toward colonialism and plantations, from distant views of the new world toward proximate, intimate understanding of distinct realms. De Bry's engravings of Le Moyne's watercolors, for example, offered visible proof of Floridian contact zones, yet they also shaped

English views of other newly "found" lands like Virginia. In particular, the engravings emphasized technologies of sight as a strategy for navigation and control of unfamiliar terrains. Like Ribault, early English explorers sought to leave visible markers of their possession of the land. On August 3, 1583, Sir Humphrey Gilbert raised the arms of England, engraved in lead and set on wood, "claiming" Newfoundland for England. Lane, in his 1584 scouting expedition to Roanoke, defied native traditions—and religious beliefs—by erecting a large, visually imposing, permanent fortress on a spiritual landscape; when he left, the natives erased all traces of it. Its disappearance, like that of the later English colonists, shows that in these vast new realms the impact of the English was minimal at first. That 115 colonists could simply disappear into such an "ocean" of forests underscores the disjunction between European "views" of the new world and sensory experiences of them.

Thomas Harriot's *A Briefe and True Report of the New Found Land of Virgina* is at the center of both approaches. While previous accounts like Amadas and Barlowes's cite both the sweetly scented allure of new world forests and the dangers that lurked within them, Harriot's narrative offers a "view" of the contact zone. In it, Virginia is a peaceful and fertile zone. De Bry's engravings of John White's watercolors emphasize this pictorially, so much so that historians often describe *A Briefe and True Report* as propaganda. Such propaganda operates by distancing readers from the contact zone through sight. Although Harriot "touches" the diversity of relations and reports, seeking to settle them "firmly," particularly in dealing with "natural inhabitants," he encourages prospective "adventurers, favorers, and well willers" for the Virginia enterprise to "see" and "know" for themselves "what the country is [*sic*]." His treatise is divided into three parts to facilitate this "more readie view." He lists Virginia's "merchantable" goods and useful commodities, citing his own experience "hauing seen and knowne more then the ordinarie," and dismissing other reports as slanderous "trifles." Taking Harriot's visual emphasis one step further, Theodor de Bry, in the 1590 edition of the text, explains that he republished Harriot's narrative with his engravings of White's images because every man should "exert himselfe for to *show* the benefits which they haue receeue of [the colony]."[50]

But Harriot's hopeful framing of such explicit, visual proof failed, spectacularly. Against such a vision unfolds the mystery of the colonists' disappearance at Roanoke. Harriot's narrative is one of the few we have of Anglo-Indian contact at Roanoke. As many critics have noted, his description of Virginia is biased toward English mercantilism, as are de Bry's engravings. By aligning English colonial desire with a mastery of technologies of sight, the narrative tells a story of English dominion of new lands, people, and especially new "merchantable" commodities. But the narrative, when read alongside White's watercolors that were turned into de Bry's engravings, also documents anxiety about the sensory realm of the contact zone, particularly anxiety about the invisible, unseen effects of bodily contact. Such anxiety focused on proximate, intimate, and olfactory contact with natives.

Harriot famously recounts native misinterpretation of English optical illusions—a lodestone, "a perspective glass whereby was shewed manie strange sightes . . . spring clocks that seem to goe of themselves"—as divine power.[51] This belief, in Harriot's view, leads the natives to declare the English once and future conquerors of the land. The English are ubiquitous, lurking in the air "without bodies," waiting to kill natives by "shooting inuisible bullets into them."[52] The phrase has perplexed literary critics and historians alike: what did Harriot mean to convey by such "invisibility"? Although his exact meaning remains a mystery, Harriot's account of native belief in invisible bullets seems to suggest that the English had an advantage due to their technologies of sight, while the natives rely on other sensory ways of knowing.

But such a conclusion reveals not only the ways vision structured English approaches to the new world, particularly at Roanoke, but also how vision has shaped our histories of contact. The implicit association of vision with science, technology, and modernity offers a view of Virginia deeply biased toward English colonial success in the new world, in spite the spectacular failure at Roanoke. Harriot's "view" supplants the historical conclusion of the colony. Though it is possible to read against the grain, there is a danger that doing so implicitly shapes scholarly approaches to the history of contact, so much so that Alan Taylor warns that "scholars have recently become so sophisticated at detecting hidden bias and covert messages that we risk writing only about the construction of illusions."[53]

Sassafras, however, was hardly a hidden or covert message in these histories. And its presence in the text (as a merchantable commodity) and in both White's watercolors and de Bry's engravings indicates the important role of olfaction in the contact zone of Roanoke. In this zone, both the English and the natives struggled to understand the novelty of the effects of contact.[54] Whereas the Roanoke imagined a forest full of unseen English spirit-bodies attacking them, the English imagined a benevolent god protecting them. As Harriot describes, a few days after the English visited Roanoke towns and villages, the natives suffered from a strange and deadly disease. The disease only affected native populations; furthermore, Harriot asserts that it only occurred in towns where there had been native resistance. Yet the experience described is unsettling, resonating with colonists' fears about airborne, environmental hazards, particularly the vulnerability of breathing in unfamiliar terrain.[55] Harriot emphasizes the novelty of this epidemic; religious men report no such occurrence for "time out of mind."[56]

The description of sassafras emphasizes the vulnerability of the English in unfamiliar terrain and the role of olfaction in navigating it successfully. Harriot's survey names sassafras as an important commodity for any potential colony at Roanoke since it was both "merchantable" in England and "most helpful" to planters in Virginia as a medical cure. Both values stemmed from its invisible, airborne qualities. Harriot notes its "most pleasant and sweet smell" and that it is "found by experience to be farre batter and of more uses than the wood which is called *Guaiacum,* or *Lignum vitæ.*" He concludes: "For the description, the manner of using and the manifolde vertues thereof, I referre you to the book of *Monardus,* translated and entitled in English, *The Joyfull newes from the West Indies.*"[57] He does not elaborate on why planters might need to rely on sassafras as a medicinal cure, sidestepping any potential reference to the pox, to English vulnerability to airborne diseases, and to Spanish and French failures in the contact zone of New Spain. Instead, he focuses on its aromatic wood. Later explorers would discover new applications for it the hard way, boiling and eating its leaves to fend off starvation.

Harriot includes the native term for sassafras, *winauk,* which suggests not only that aromatics were a subject discussed in early Anglo-Indian exchanges but also that this particular one was plentiful. Unlike vision, olfaction offered proximate, and sometimes inexact, knowledge of these new realms. Explorers often falsely reported the presence of botanical matter. Orris root, for example, a popular aromatic ingredient used to scent cloth, hair, and tooth powders, was routinely reported as common to the new world, despite that it is not native to North America.[58] Smell offered proof of a commodity's presence and potential value, even when unseen. Harriot includes civet cats in his list of merchantable goods, claiming that "good profites will rise" from them. Though no colonist saw a cat, he remained convinced of their presence. Colonists noted a distinctive smell, left "where one or more had lately been." The presence of civet cats was also corroborated "by relation of the people there are some in this country."[59] Similarly, Harriot suspected that there were many "sweet gummes and many other Apothecary drugs" in the area. But, without an expert, he could only hint at their usefulness.[60] He reports confidently, however, that Virginian winauk is the same described by Monardes, suggesting how native olfactory mapping of local terrain supplemented English visual ones.

Like the native echo of *winauk* resonating in Harriot's texts, de Bry's images contain their own phenomenological trace of the contact zone: John White's watercolor sketches. Though John White's watercolors supplement Harriot's narrative, they also function as a personal archive of sensation. As such, they participate in what Mary Fuller has described as the ways personal memory becomes historical memory. Sassafras was an important part of this process. Sent to Virginia to assist Harriot in mapping and cataloging the Virginian coast, John White was a gentleman limner.[61] He kept detailed notebooks of his journey and made copious sketches of native portraits, scenes of native life, details of flora and fauna, and intricate maps. Though most of these notebooks were lost, and the catalog of watercolors inspired from them were shown only to a very limited selection of people, de Bry's reproduction of them in his *America* series made them some of the most influential and widely disseminated images of the new world.

White's perspective is hardly unique (his watercolors reflect an amalgamation of travel narratives describing flora, fauna, and peoples from Turkey, Florida, Brazil, Newfoundland, and Virginia) or objective (they also reflect wildly optimistic view of temperate climate, peaceful people, and bountiful crops). Yet it reveals a struggle to objectify his firsthand experience of the new world and present it to others, namely Raleigh. That these images survived even after Roanoke colony disappeared emphasizes the ways subjective, phenomenological engagements in the contact zone could become spectacular and iconic representations of English colonialism.

Risking his own family and money, White invested everything in the success of English plantations in Roanoke. His granddaughter, Virginia Dare, was the first English child born in the new world; she remained in Roanoke with her parents, waiting with the other colonists for his return. Furthermore, he was a gifted limner, with powerful connections to Harriot, and through him to Raleigh and the court; White traveled to Virgina six times in ten years, mostly to search for lost colonists. All of this, Kim Sloan suggests, "situates him perfectly as our Elizabethan interpreter of the New World."[62] His phenomenological engagement with Roanoke remains one of the more compelling aspects of its archive. How did Roanoke smell? What did it sound like? How did it feel? Do his visual images sketched in the field betray the broader sensory worlds in which they were engaged or do they merely reflect illusory desires of the metropole?

Much has been made of the optimism that opened White's career and shaped his depictions of the Anglo-Indian contact zone. In his watercolors, he greatly extends the Virginia growing season (planting three successive crops), and his natives generously offer foodstuffs (despite that White and his colleagues arrived in the midst of a terrible drought, when natives were most likely anxious to preserve holds from strangers).[63] Less attention has been paid to his disillusionment and his framing of it in sensory terms. His final letter to Raleigh, before retiring to Ireland, expresses his own disappointment in the venture: "I would to God it had been as prosperous to all, as *noysome* to the planters" (emphasis added).[64] His use of *noisome* had olfactory connotations: it signaled not only annoying, disagreeable, or troubling hazards but also, increasingly in the sixteenth century, stench.[65] What was once so sweetly alluring has become cloying. White's disappointment is palpable; he describes himself as "lucklesse" and "froward" in his early hopes.[66] Though he concludes that he has been content with the "sight" of the new world, his concern for the planter's noisome struggles gesture toward other sensory experiences in the new world, experiences both good and bad that could not easily be represented in travel narratives.

De Bry embellishes the environmental backgrounds but retains White's focus on the smiling, naked natives, who function as important visual clues to wary investors in the Virginia plantations. Their scantily clad bodies signal a temperate (and potentially fruitful) climate, and their gestures convey their status as helpful (and hopefully peaceful) native neighbors. Against a stark white background, in White's original drawings, the figures of the natives and individual details—tattoos, jewelry, hairstyles, clothing, postures, and gestures—are clearly visible. If such details signal an illustrious future for English plantations in Harriot's published treatise, then they also point toward a not-so-distant past: the artistic moment of production, when John White was in the field sketching. As such, they function as an archive of the ecology of contact and, perhaps, a trace of White's phenomenological experience of it.

Consider, for example, White's depiction of "a festive danse," which de Bry engraves and includes under the label "their danses which they use att theigr hyghe feastes." Harriot's narrative recounts a great and solemn feast that occurs at a "certayne tyme of the yere," "in a broad playne," "after the sunne is sett," and which included neighbors from nearby towns. In this feast, men, "attired in the most strange fashion they can deuise," dance and sing around three "fayre virgins," who embrace in the center, "abowt which are planted in the grownde certayne posts caured with heads like to the faces of Nonnes couered with theyr vayles."[67] So important is this feast that White includes two images of it: one in detail and one placed on the map of the Secotan village.

In his watercolor of the town of Secotan, White offers a snapshot of life there. Along with images of their tomb, their place of solemn prayer, and their fields of

John White, "A Festive Danse," watercolor over black lead (circa 1584). Scholars have emphasized the role of corn and tobacco in this painting, though the branches most natives hold strongly resemble sassafras. © Trustees of the British Museum

"ripe corn," "green corn," and "corn newly sprung," the dance is an integral part of the landscape. White describes it as "a ceremony in their prayers with strange iestures and songs dansing abowt posts carued on the topps lyke mens faces."[68] The dancing men hold rattles while others rest, waiting to join the dance, or guard the corn crops in the distance. In his detail of the festive dance, White seeks to represent the dance as motion, elaborating on such strange gestures. Each figure is surrounded by light shadowing, which demonstrates the shape of the dance. White also includes light shadows of branches of leaves, which seem to be passed to each participant as he enters the circle. These shadowy branches resemble sassafras leaves. It is as if White seeks to represent not only the figures of the body but also the sensory effects of their ritualized motions. The smell of sassafras is captured by representing the branches in motion. Given that the dance occurred after sunset, such a scent would undoubtedly be an important component of the colonists' experience of it.

The men are painted holding rattles and wearing feathers and pouches. Some scholars argue that both represent tobacco, which Harriot describes as an important part of native rituals (although Harriot mentions only powdered ceremonial uses).[69] Indeed, de Bry's engravings interpolate tobacco and corn into these scenes. He adds fields of tobacco (along with giant sunflowers) to the depiction of the Secotan town, and though he eliminates the shadowy steps of the dance, de Bry emphasizes the use of branches in the dance, filling them in and supplementing them in with husks of corn. These embellishments are consistent with others he makes; he translates White's blank backgrounds into specific environments, even if they are generic. The effect is to translate White's sensory record into a discrete narrative. The dancers are frozen in time, and the sassafras no longer represents a smell but a visual symbol of future monetary success. Both White's and DeBry's representations of Secotan town life purport to offer an "objectification" of sensory data, rendered through an imagined aerial perspective.[70] But as critics note, there are hints of a counternarrative embedded within them. John White's watercolor of the dance, with its shadowy aromatic branches and vivid movement, are one such clue, suggesting that the English were just as captivated with the natives' rituals, and invisible, aromatic technologies, as the natives purportedly were with theirs.

The English settlement at Roanoke, like Gilbert's journey, was a spectacular failure. By 1586, the natives and English were involved in armed struggles. The natives had the distinct advantage of stealth, whereas the English were armed with matchlock muskets, which required that one strike a dry match and ignite the gunpowder to fire the weapon. The scent of a struck match alerted the natives hiding in trees, armed with bows and arrows, of their presence. Though the guns were loud and smoky, they were ineffectual against the natives' guerrilla tactics. The English built a large, visibly imposing, permanent fort on the settlement. The natives retaliated by withdrawing their support, allowing the English to starve. Lane reported to English investors that he and his men survived by eating sassafras leaves alone, which again testifies to the ubiquity of the plant. When Lane and his men returned to England in 1586, the natives destroyed all evidence of an English settlement. White, along with 150 colonists, returned to Roanoke in 1587, rebuilding the abandoned and destroyed settlement. Conditions soon worsened and White returned to England to secure additional supplies, leaving approximately 115 colonists behind. When he returned in 1590, all traces of them and their settlement had vanished into the thick, dense forest air.

The disturbing disappearance of the colony presented Raleigh with a challenge, not unlike Gilbert's twenty years prior: how to inspire English colonists in the wake of such tragic news. Sassafras, as both a luxury perfume commodity and as a potential medicinal cure, was one possibility. Its value in Europe provided an

Theodor de Bry, after John White, "Their dances which they use at their high feasts" (1590). DeBry adds long, narrow tobacco leaves and feathery leaves of corn to White's depiction. © The British Library Board, shelfmark C.38.i.18. All rights reserved 06/08/2010

incentive to mitigate the failures at Roanoke. In sixteenth-century accounts of the contact zone, sassafras served as an important symbol of potential English success in the new world, despite that phenomenological descriptions of its smell in those same accounts underscore English vulnerability to native attack and hostile environments. By the start of the seventeenth century, however, it also functioned as a symbol of the failure at Roanoke, its ephemeral scent in English markets a troubling reminder of what remained illusive. In published accounts of return voyages, Raleigh cites greed as the reason the colony remained lost; rather than searching for the colonists, the voyagers collected sassafras instead.

As Raleigh's men continued to search the Carolina and Virginia coasts for signs of the settlers, Sir Humphrey Gilbert's nephew, Captain Bartholomew Gilbert, along with Captain Bartholomew Gosnold, sailed (without Raleigh's knowledge) to the "north part of Virginia," a region now known as New England.

Gosnold's and Gilbert's voyage was strategic, short, and profitable, if not entirely successful. Planning to retrace Giovanni da Verrazzano's route to Narragansett Bay, Gilbert and Gosnold sailed west to Maine, down the coast toward Provincetown, around Cape Cod, and into Buzzard's Bay, near Martha's Vineyard. There, they established a small fort on the western part of Elizabeth Island and commenced trading with the Wampanoag people. They had hoped that such a settlement would be permanent; Gosnold was to remain in command of the settlement while Gilbert navigated the *Concord* back to England. Due to a dramatic miscalculation in supplies, in which the proposed settlers would be left with six weeks' worth of supplies rather than six months' worth, the settlement plan was abandoned. The *Concord* returned home to Bristol with her entire crew and 2,200 pounds of sassafras, which was almost a ton.[71]

Raleigh discovered Gilbert and Gosnold's endeavor surreptitiously, only after their profitable cargo saturated England's sassafras market.[72] Furious, he worked to impound the *Concord* and confiscate their remaining cargo; he even sent Harriot from London to Southampton to seize a cartload of sassafras that had been sent ahead.[73] Raleigh's goals for such seizures were twofold: protect his own financial interests in sassafras and secure his charter. He addressed his financial interests first, arguing that any additional importation of sassafras would "cloy" England's markets.[74] Raleigh quickly recognized, however, that he could parlay their success into his. Recasting their voyage as part of his new world endeavors, he published an account of the voyage alongside of reports of his searches for the lost colony.

Repackaged as "Northern Virginia" rather than harsh Newfoundland, the north again seemed economically promising. Gilbert and Gosnold's lucrative journey revived English colonial endeavors, and their haul of sassafras helped to transform the failures of Newfoundland and Roanoke into a promising future that would coalesce around a different "Virginia." John Brereton's widely read *A Briefe and True Relation of the Discoverie of the North Part of Virginia*, published in 1602, represents this new approach, suturing together Brereton's account of Gilbert and Gosnold's journey and a brief description of Captain Samuel Mace's search for the lost colonists of Roanoke.[75] In combining these two distinct journeys, Brereton conflates Gilbert and Gosnold's sassafras collection in the north in 1602 with Captain Mace's failure to find the lost colony of Roanoke in the south the same year. Mace fails to find the colonists because of his greed; he searches for sassafras rather than colonists.

Perfume was an important part of this political repackaging, its ephemeral scent linked to the lost colony of Roanoke. In Brereton's account, Mace chased the wrong scents. Brereton reports that Mace was "a very sufficient Mariner, an honest sober man, who had been at Virginia twice before," suggesting his knowledge of the terrain. In this way, Mace was like others that had searched for the lost colony. As Brereton notes, Raleigh "hath sent fiue seuerall times at his owne charges," concluding that "some of them following their own profit elsewhere; others returning with friuolous allegations."[76] He then reports that the ship returned loaded only with lucrative scent ingredients. The "brief note" on the voyage disregards the difficult navigation conditions along Cape Hatteras, blames Mace for failing to find the colonists, and catalogs the tremendous merchandise available for harvesting.[77] That sassafras could conflate disparate journeys to North Carolina and New England demonstrates its important role and the role of olfaction in shaping the representational history of Anglo-Indian contact.

Brereton's condemnation of Mace was political. Gabriel Archer's account of the same journey, published in 1625 after the success of Jamestown, reports the experience of collecting sassafras in the north very differently from Brereton, particularly the phenomenological experience of vulnerability and fear experienced by the English in these zones. Archer was a chaplain on Gosnold and Gilbert's voyage. Like Brereton's, his narrative highlights the prosperity of the voyage from the start. Even the multiple signs of landfall were portentous: the water turned to a "yellowish greene," the ground became "sandie" with "glittering stones . . . which might promise some Minerall matter," weeds and woods floated by, and finally, "we had a smelling of the shore, such as the from the Southerne Cape and Andaluazia in Spaine."[78] The natives were comely, tall, and surprisingly adept at trade, undoubtedly due to previous participation in the fishing trade to the north. Newfoundland, in this account, is hardly a failed encounter in an empty, harsh landscape but rather an important first step in Anglo-Indian contact. Its history emerges through native familiarity with "Christian" language. As a result, in Archer's view, the natives were altogether more knowledgeable about the English than the English were of them.[79] Brereton corroborates Archer's conclusion, but de-emphasizes its implications, noting only that "they pronounce our language with great facilitie."[80]

Brereton's narrative describes the landfall at Maine as an encounter with a veritable paradise. The environment is so lush that the sailors, "when coming ashore . . . stood a while like men rauished at the beautie and delicacie of this sweet soile," marveling that "nature would shew herself aboue her power artificiall."[81] Despite such a spectacular setting, Brereton remains focused on profits, noting that sassafras trees are "great plenty all the Island ouer" and that they are "a tree of high price and profit."[82] The first listing, and the only one to include

European values, is "Sassafras trees, the roots whereof at 3 s. the pound are 336.1 the tunne."[83] Archer's account offers a different perspective, narrating difficult labor and harsh working conditions. Though the trees were plentiful, the labor of collecting sassafras was difficult. For sixteen days, the labor of collecting sassafras and of building a permanent settlement consumed the crew, interrupted only by native encounters or foul weather.[84] Archer's narrative also documents that the Micmac, Massachusetts, and Wampanoag offered assistance with the collection of sassafras, perhaps using this as an opportunity to examine the English and their defenses.[85]

Though he emphasized the ease with which sassafras might be culled from the landscape, Brereton also claimed that England's sassafras market was limited, suggesting that any subsequent voyages and attempts to harvest such merchandise would need supervision under Raleigh's charter.[86] Though it is tempting to trust Raleigh's conservative and self-interested assessment of Europe's sassafras market, sassafras continued to buoy English investments in the new world, culminating in the Virginia Company's stakes at Jamestown. Despite Raleigh's warnings about a glut in the market, the success of Gilbert and Gosnold's voyage, along with the popularity of Brereton's narrative, increased demand for sassafras. Raleigh himself chartered two more voyages to the Chesapeake region to search for it. These voyages were important precursors to the permanent settlement at Jamestown, sharpening the sensory lessons learned at Roanoke and providing explorers with opportunities to apply them to new realms.[87]

For example, Martin Pring voyaged to the north in 1603 in hope of emulating Gilbert and Gosnold's haul. However, his account documents how quickly Anglo-Indian relations evolved. Like Archer's narrative, Pring's journal was published in 1623, almost twenty years after the voyage itself, as part of promotional materials for Virginia. It also describes contentious confrontations with the natives, linked to the labor of collecting sassafras.[88] Over two hundred natives overwhelm the small group, who were napping after an arduous morning of labor. The natives attacked the sleeping sailors, who awoke in time to defend themselves thanks to two mastiffs. The natives later returned and set fire to the surrounding forest, chasing the English from the area and making any subsequent return trips for sassafras impossible. Such tactics undoubtedly sought to attack the profits that drove the English there in the first place. Pring's account thus documents that the natives had learned from previous encounters with English sassafras seekers. The English were learning, too; native tactics described here, in particular the use of fire as an olfactory menace and a tool of environmental control, are very similar to ones the English later employed at Jamestown.

Sensory Substitutions: Jamestown, Tobacco, and the Colony of Virginia

After reading early accounts of Amadas and Barlowes's early reports of the Virginia coast, and perhaps inspired by the impending departure of the Jamestown expedition, Michael Drayton wrote his ode "To the Virginian Voyage" (1606). In it, he commands the brave to "go and subdue" the newfound land. Virginia is a paradise, where "nature hath in store Fowl, venison, and fish / And the fruitful'st soil, / Without your toil, / three harvests more."[89] Chief among these resources are cedar, cypress, pine, and sassafras. Indeed, sassafras was written into the charter for Jamestown. The irony, of course, is that Virginia's success depended not on these fragrant resources but rather on a different kind of crop: tobacco.

In his analysis of the sensory worlds of early America, Peter Hoffer argues that when the English renewed their attempts to colonize southern Virginia, they applied the lessons learned at Roanoke, namely the importance of sensory manipulation of the environment.[90] Though the first two decades of the plantation would be marked by crises, by the late 1620s, the English secured a permanent settlement in the new world through their success at what Hoffer terms "sensory imperialism."[91] Tobacco was a large part of such imperialism. The successful planting of tobacco crops after 1622 enabled Jamestown to emerge as a powerful and stable settlement and allowed the English to forever alter the landscape of Virginia.

If tobacco marks the English's success in Virginia, then sassafras represents its failures. Tobacco defined the English presence in the new world; as Hoffer concludes, "the plough was the pen with which the English husbandman wrote his name on the land."[92] Sassafras, however, could only be collected; it could not be cultivated. Any English presence associated with it was ephemeral as its scent. Since its roots were the source of its value, a sassafras tree, once collected, was forever destroyed. Thus, the colonists' early survival in Jamestown depended largely on sassafras's overwhelmingly abundant presence in the surrounding areas; settlers could collect it even while under attack from native forces. In 1610, sassafras's fragrant wood was one of the first commodities Smith sent back to the Virginia Company as proof of the settlement's viability. Yet sassafras was a commodity better suited to raiding missions than to permanent settlements.[93] As the Jamestown settlers attempted a more permanent presence on the landscape, they destroyed the surrounding stores of sassafras. By 1622, the settlement failed to make its quota of 30 pounds of sassafras, and the Virginia Company penalized tobacco planters for failing to meet their share.

Though such deforestation undoubtedly left an impact on the landscape, it was nowhere near the scale of tobacco plantations. And, despite that the English used sassafras to survive those early hardships, sassafras remained resoundingly under native control. English struggles to permanently inhabit Virginia emphasize this point. The territory to the north of the valley of Powhatan was known as Winauk or Wynaoke, a term that connoted both "roundabout way" and, as Harriot noted in his *Briefe and True Report,* land of sassafras.[94] Its name reverberated with the English experience of the space and undoubtedly led the English to equate the landscape with odoriferous qualities of native realms.[95]

Thus, a wholesale shift in the landscape, both in terms of its name and its sensory worlds, was required for the English to inhabit such a space. To do so, they would revive Humphrey Gilbert's dichotomous distinction between sensuous, brute animality and English reason, locating sensory experiences of space— olfactory ones in particular—firmly in the former. Whereas up to this moment, English success in the new world required olfaction, after 1622, and what Peter Hoffer terms John Smith's successful "sensory warfare" against Opechancanough and the Powhatan confederacy, the smell of the contact zone was associated with natives' brute physicality rather than English humanity. In 1622, the Jamestown settlers, struggling to meet their sassafras quotas while planting fledgling crops of tobacco, suffered a vicious attack from the Powhatan confederacy. Peacefully entering the homes of the English unarmed, the natives used English tools as weapons against them (and their animals). A quarter of the settlers were tortured, killed, decapitated, and hung in plain sight; such violence was meant to demonstrate native control over the realm, even those spaces purportedly under English control.

The English, however, answered such violence with sensory warfare tactics described in accounts like Pring's and Archer's. Rather than attack the natives directly, the English sought to change the landscape in which they dwelt. For Governor Francis Wyatt, writing to defend his leadership of the colony in the wake of the attack, the solution lay in deforestation. The English must seat "the whole colony upon the forest." Deforestation would enable the landscape to become more hospitable for the English; it would make the climate drier, warmer, louder, less diseased (due to fewer mosquitoes), and more arable for tobacco planters and their animals. John Smith, of course, had called for such measures from the beginning, particularly for "carpenters, husbandmen, gardeners, fishermen, blacksmiths, masons, and diggers up of trees, roots well provided, than a thousand of such we have."[96]

Though his first encounter with the Powhatan Valley was characterized through olfaction, Smith quickly derided the natives through their association with smell. Sassafras, and its smell, was a profit to be harvested from the landscape, but the natives would have to be evacuated from it. Later, Smith derided the natives as mere "brutes, vile and stinking" in their customs," despite his noting of the foul stench of the English and their deplorable living conditions.[97] His description of the natives as stinking "brutes" demonstrates how smell participated in the violence of early Anglo-Indian contact. Preserving the role of olfaction in the terrain of the new world by associating it with English perception of bodies rather than landscapes, Smith transforms the natives, and their brute physicality, into objects, rather than subjects, who smell.

The English's success in Virginia ultimately depended on visual mastery of new landscapes. With the successful cultivation of tobacco, the English were able to rework the landscape so that it mimicked the sensory worlds of England. Such a distinction between subjects and objects and between bodies and environments was never as stable as the English hoped, for the nose prominently displayed how ephemeral contact with other sensory worlds could permanently reshape bodies. In her reading of how technology and science shaped the Anglo-Indian contact, Joyce Chaplin begins with two tales of diseased noses—one native, one English. Both document how the nose, as an organ of smell, represented "the tip of the problem" in navigating the sensory worlds of the contact zone. The first, a Santee tale told in the Carolinas in the early 1700s, relates two natives who emerge noseless from a remote forest, having survived a terrible illness. When asked about the loss of their noses, they reply that they had conversed with the "Great Being" who "had promis'd to make their Capacities equal with the white People in making Guns, Ammunition &c." in return for their noses.[98] In such a tale, the natives demonstrate a profoundly sardonic awareness at how the history of contact would be told: sagacious, native knowledge would succumb to English weaponry. As Chaplin argues, noses represented the "tip of the problem," if one seeks to understand why "the inhabitants of colonial America, Indian and English, dwelled so obsessively, and ultimately, so decisively on bodies and technologies."[99]

The Santee story demonstrates how olfaction bears the cultural trace of a brief moment in which the environment of the contact zone absorbed, affected, and transformed the bodies within it. A story about Thomas Harriot tells the story the other way round. After returning to England, Harriot, mathematician, famous surveyor of Virginia, and enthusiastic proponent of smoking, suffered from an ulcer on his face, which slowly consumed his entire nose and ultimately killed him. It was most likely a nicotine-induced cancer. Harriot's colleague, mocking his mathematical belief in atomism, described his death as the ultimate "nihilium" or triumph of nothingness: "for in the top of his nose came a little red speck, which grew bigger and bigger and at last killed him."[100] Though the joke pokes fun at Harriot's mathematical beliefs, it gains teeth through its invocation of deadly nothings, whether they be atoms, tobacco smoke, invisible bullets, or exogamous disease. Harriot's ghastly and painful condition would have registered as a symptom of the pox, a pox conflated with the effects of the new world on English bodies, which made visible how vulnerable English bodies were in new world environs.

Though artificial noses could mitigate the cosmetic effects of the disease, such prosthetics, usually made of silver or of grafted skin, in some ways functioned like sassafras: an important visible sign of the invisible hazards of breathing dangerous air. Prosthetic noses, after all, hindered one's ability to smell. They functioned as brute bodily material rather than as organs of sense. Prosthetic noses also had visual associations with plague beaks used in continental Europe. Those elongated beaks, stuffed to the brim with potent herbs and spices, isolated the nose as a very vulnerable sensory organ in diseased spaces. As the next chapter demonstrates, olfaction's association with transmission of disease would galvanize new kinds of fears about bodily contact, especially with those ravaged by plague in urban spaces.

CHAPTER FOUR

Smelling Disease

Rosemary, Pomanders, Shut-in Households

Despite its title, Thomas Dekker's *The Wonderfull Yeare* suggests that 1603 was anything but. After a description of the funeral of Elizabeth I, the famous plague pamphlet provides a "picture of London, lying sicke of the plague."[1] It is a gruesome picture, to be sure. The outbreak of the plague in 1603 was virulent and swift; in a few short months, more than thirty thousand Londoners were dead. Almost all the rest of the city's inhabitants, Dekker's pamphlet suggests, were in mourning or exile. Yet the city remained crowded with the ephemeral traces of those who had departed. The text begins its description of the plague by inviting the city's ghosts; grieving "desolate hand-wringing widows"; "wofully distracted," "disheueld" mothers; "out-cast and downtroden Orphanes"; and those "genii" who escaped to the Antipodes to "cast a ring" about the author so that the sensory horrors of the plague might be captured in prose—so that their groans might "echo" in his pen and their tears "rain" into ink.[2]

Sources like Dekker's *A Wonderful Yeare* suggest that the sensory world of London in 1603 was ghastly. Despite its exponential population growth, London was literally a ghost town; those who remained in its midst shut themselves into their homes, in an attempt to shut out the horrors, including nights spent like one in "a vast, silent Charnell-house."[3] Such emptiness did little to relieve the fears of those who remained in the city, or near it: the smell of decaying bodies made the scale of London's plague palpable in ways previously unimaginable.[4] The stench invoked the ephemeral, sensory traces of the dead and their bodily remains. Though commentators proffered a wide variety of theories about the cause of the plague (Dekker himself rejected the theory of infected breath in another pamphlet), most, when confronted with the material effects of the disease, emphasized the dangers of smelling noisome air.[5] As one commentator notes, the plague was dangerous and airborne, composed of "a rooten and corrupt ayre by a hidden and secret properties which it hath."[6]

This chapter explores how such "hidden and secret" properties materialized through smell. Though the exact nature of such properties was up for debate, the presence of any strong smell was a reminder of how the body was vulnerable to its environment and to unseen influences circulating within it. In such zones, olfaction was a key way that embodiment was experienced. As I argued in the previous chapter, within certain environments perceived to be dangerous (like the contact zone), olfaction was an important tool of navigation. Yet it rendered the body vulnerable to its proximate environment. In such spaces, sniffing even the sweet scent of sassafras was a dangerous activity (even though, at home, it represented the allure of English prosperity in the new world). Yet one could hardly stop breathing. As the jokes made about Sir Thomas Harriot's gruesome illness and death (due to what was most likely a tobacco-related cancer of the nose) revealed, the cost of such exposure was as plain as the nose on one's face, or the lack thereof.

Anxiety about bodily exposure to "dangerous" air intensified during outbreaks of plague. As the plague became associated with particular urban zones, and, more broadly, with the metropolis as a whole, panic over how and when it might spread beyond these locations increasingly focused on the invisible networks that connected them. The smells in the air showed that there was virtually no way to police such boundaries or, more importantly, to protect the porous body from infected zones. Ensuing fears brought London's festive life to a screeching halt. Public playhouses were particularly dangerous, since the worst-affected parts of the city were its fringes, where many of the city's amphitheatres were located.[7] Even processional celebrations of James I's coronation were delayed by proclamation of the king for fear of spreading the plague to the countryside.[8] And the king proceeded from the tower to his coronation in Westminster entirely by water in order to lessen his own threat of exposure. Suffering from the epidemic, the city was described as a dank, crowded, noisome space of death; in Dekker's text, slow-burning lamps illuminate dark hollow corners of the "diseased Citie," full of "half mouldred" corpses. Its streets are "strewde with blasted rosemary, fatall Cipresse and Rue, thickly mingled with heaps of dead mens bones," filling a passerby's nostrils "with noysome stench."[9] Dekker complained that the cost of herbs and garlands increased in London markets, with rosemary jumping from "12 pence an armefull" to "six shillings a handfull" as Londoners attempted to protect themselves from diseased air. Describing the city's inhabitants as "shrinking" their heads into their collars, "stuffing" their noses with wormwood and rue, and sticking themselves with so many branches of rosemary that they look like cooked boars for Christmas dinner, Dekker evocatively captures how fears about diseased air altered previously unconscious re-flexive acts such as breathing.[10] This aggregation of scents emphasized the poignant atmosphere of death and decay.

The Smell of Rosemary and Rituals of Memory

The particular smell of rosemary was a complicated signifier of disease and death. Commonly associated with rituals of memory, its scent symbolized betrothals, marriages, and funerals. Yet the scale with which it was used as a plague preventative during the outbreaks of 1603, 1609, and 1625 suffused it with new kinds of signification, specifically in regard to olfactory encounters with disease in crowded urban spaces. The scent of rosemary thus became a harbinger both of erotic, affective promise and diseased peril. Its presence in Dekker's text em-phasizes this olfactory paradox: the noisome horrors that torment London during its plague are juxtaposed with the promise of a verdant spring. Dekker's text opens with an invocation of dynamic, seasonal renewal. In January, Vertumnus makes "lustie progresse . . . perfuming all the wayes that he went, with the sweete Odours that breathed from flowers; hearbes and trees."[11] By the virtue of such "excellent aires," the spring sky had a clear complexion, without a single wart. The sun, appareled in gold like a "bridegrome" carried not "gilded Rosemary" but rather the gilded horns of March; spring was poised to pastorally unfold, complete with frisky lambs, leaping goats, piping shepherds, and lovers who "made Son-nets for their Lasses, whilest they made Garlands for their Louers."[12] As the country was "frolike, so was the Citie mery." All of this quickly changed with the death of Queen Elizabeth and the outbreak of the plague. When rosemary ap-pears in the pamplet, it marks not spring but death. Only the promise of a new king dispels the gloom: James I appears as a radiant sun from the north, "whose glorious beames (like a fan) dispersed all thick and contagious clowdes."[13]

While it is easy to read such a description symbolically, many of James I's coronation pageants in 1603 materialized such displays of royal power over the blighted city. Whereas Tudor royal power had concentrated itself through the visual beauty and the powerful perfume of the damask rose, Stuart power focused on environmental control and initiatives of public health. This was reflected in James I's reincorporation of London's many mysteries (particularly those that traded in perfume ingredients, as I argue in the next chapter). It also shaped the elaborate displays that were part of the king's coronation pageants, which oc-curred after the city had suffered tremendous losses from the plague. As Dekker notes in his *Magnificent Entertainments*, the pageants included a citation of one of Henry VIII's famous court performances involving rosewater, now mounted on a civic scale. According to Hall's *Chronicle,* Henry VIII included an aromatic fountain in a royal interlude in 1527; from its summit "was a fayre lady out of whose bresetes ran aboundantly water of mervueilous delicious saver," which was undoubtedly damask rosewater.[14] A similar fountain was created in 1603 and placed at Sopers Lane, an area known for its medieval associations with the city's spicers, during James I's coronation pageant.[15] The fountain had five mounts— one for each of the five senses—that flowed with milk, wine, and scented balms. According to Dekker, together they allegorically represented James I's ascent as a Phoenix rising from Arabia's spiced nest. But they also materialized his royal ascent over the blighted city: described as being "languorous" with grief until James's arrival, they "sprung" to life in his presence, embodying royal magnificence through the sensory delights of red wine, white milk, and perfume.[16] To citizens besieged by plague, such perfumed benevolence was undoubtedly a welcome relief though not without its own risk.

London's diseased smellscape described in Dekker's *Wonderfull Yeare* (along with James I's perfumed relief in his *Magnificent Entertainments*) reveals how phenomenological encounters with death shaped collective urban experience. Belief in the ability of airborne odors to transform an environment was built on just such experiences, yet there was profound confusion over transmission. Was noisome air itself contagious? Did bad smells breed the plague? Or did foul air merely intensify the experience of disease? Representations suggest a host of fears about the relationship between scented air and undetectable traces of contagion. Consider Shakespeare's *Henry V* and the king's curious addendum to the play's rousing "St. Crispian's day" speech.[17] In this famous oration, Henry inspires his troops by imagining his small army's collective bodily power: "we few, we happy few, we band of brothers" (4.3.60).[18] Outnumbered by French troops five to one, Henry imaginatively memorializes his soldiers' valor as a cohesive unit. Yoking the bloodshed of warfare with heredity, Henry unites his men through metaphors of the body: the army becomes a band of brothers, "[f]or he to-day that sheds his blood with me shall be my brother" (4.3.61–62). But, as his second speech acknowledges, the violence of war will mingle his soldiers' bodies, creating a powerful—and gruesome—odoriferous hazard: "A many of our bodies shall no doubt / Find native graves . . . / And those that leave their valiant bones in France, / Dying like men, though buried in your dunghills, / They shall be fam'd; for there the sun shall greet them, / And draw their honours reeking up to heaven; / Leaving their earthly parts to choke your clime, / The smell whereof shall breed a plague in France" (4.3.97–105).

The violence of warfare is perpetual and atomized; the bodies and bones of Henry's soldiers, buried in dung, "breed" together an airborne, odiferous "plague" that "chokes" the "clime" of France with its smell of dung and the soldiers' rotting corpses. Henry's speech emphasizes that sensory perception of a foul stench was dangerous both in overt and in more subtle ways. On the one hand, the stench itself is a plague. On the other hand, the plague is also bred from the atomized, rotting bodies of the soldiers, demonstrating profound anxiety about a body's vulnerability to its environment and to unseen, imperceptible material circulating within it, including atomized, diseased bodily matter. For this reason, strong scents of any kind—both pleasant and not—were cause for great concern in early modern locales.

Though Shakespeare's play predates the outbreak of the 1603 plague, it demonstrates how death was conflated with rotting smell, smells capable of breeding new kinds of harms. In 1603, Thomas Lodge echoed such a belief, describing how the plague "proceedeth from the venemous corruption of the humors and spirits of the body, infected by the attraction of corrupted aire."[19] By 1625, the only cure for the plague was completely altering one's environment; because "sicke bodies infect the outward Aire, and that Aire again infects other Bodies," one physician advised those at risk to "[f]ly with speed from the infected place, lest by a little lingering that infection (which you would leave behind you) goe along with you . . . and flie not a little way, but many miles of, whither there is no probabilitie of trading, or recourse of people from the place forsaken: and where there are high hills betwixt you and the infected coast."[20] Spatial distance—preferably with natural barriers like hills—severs proximity to afflicted bodies, linking disease to the inhalation of air tainted with others.

Yet death was also conflated with the sweet scent of rosemary, a smell previously associated with erotic love and fidelity. In Shakespeare's *Romeo and Juliet,* rosemary is an erotic tool of memory, linking young love with its potential triumph over death. Commenting on young Romeo's passion for Juliet, the Nurse remarks, "Doth not Romeo and Rosemary begin / both with a letter?" (2.3.188–90).[21] Rosemary's use as an olfactory trigger for personal memory made it a powerful—and complicated—scent when it became associated with collective memories of shared urban experiences of the plague. Under these conditions, rosemary and sweet waters—once a cultural signifier of elaborate bridal rituals—marked instead the sinister, material presence of disease. Dekker, too, underscores this transformation when he writes of "a husband and a widdower, yet neuer knew his wife: she was his owne, yet he had her not: she had him, yet neuer enioyed him: heere is a strange alteration, for the rosemary that was washt in sweete water to set out the Bridall, is now wet in teares to furnish her burial."[22]

Likewise, in Thomas Middleton and William Rowley's *A Faire Quarrell* (1617), rosemary is a stage property used to materialize both meanings on stage. In the play, Jane, secretly wed to Fitzallen, is pregnant with her husband's child, yet she is arranged to be married to another—a foolish Cornish gentleman, Chough. Her physician discovers this and uses the knowledge to compel sexual relations with her, lest he tell Chough about Fitzallen. The drama comes to a head on her wedding day. Trimtram, Chough's servant, interrupts the physician's advances, carrying rosemary to be used later in the day during the wedding service. Denied sexual favors, the physician seeks revenge, inquiring about Chough's plans to be married. Trimtram replies that his master must be married that day, "else all this rosemary's lost" (5.1.40).[23] On discovering Jane's inconstancy, Chough finds another use for the rosemary, vowing to burn "all the rosemary to sweeten the house, for in my conscience 'tis infected" (5.1.121). Trimtram, ever concerned about waste, cautions that there will be "charges for new rosemary else" since "these will be withered tomorrow." Chough cares not, commanding him to "make a bonfire on't to sweeten Rosemary lane." Rosemary Lane, just north of the tower, was located in the suburbs of London in St. Botolph's Parish-Aldgate, a place hit hard by both the 1603, 1609, and 1625 outbreaks of the plague.[24] The parish's cemetery was located on Rosemary Lane and thus was a space in which rosemary would also have been employed as a prophylactic fume.[25] The play's associations link the smell of rosemary to sexual inconstancy, "infected" domestic interiors, and public bonfires used to combat the plague. Most audience members would have experienced such smells firsthand, linking the social and olfactory worlds of the play with those of plagued London.

Plague, Confinement, and Perfume Prophylactics

Early modern medical authorities recognized perfume as both a contagion and a cure for the plague, marking and masking the presence of dangerous air. Within early modern medicine's conceptions of olfaction, breathing and smelling were implicitly linked. Foul air and the fragrances used to fight it were particularly worrisome for they materialized the ways a body was inextricably linked to its environment. Consider, for example, Cloten's entrance in Shakespeare's *Cymbeline* (1609), a play linked to Stuart iconography (and one that debuted during a plague year in London).[26] Covered in the blood of another, Cloten is described as "reeking as a sacrifice," and counseled to "shift" his shirt because "where air comes out, air comes in" (1.2.1–3). Punning on Cloten's scent, an obsequious courtier puns that no scent "abroad" is as "wholesome as that you vent" (1.2.4). Taken literally, the courtier's pun posits that Cloten's bleeding body is vulnerable to the air surrounding it: one could exchange rank bodily vapors for fouler environmental scents, particularly in sick environments.

The widespread use of pomanders, womb-fumigators, and perfumed pessaries as plague remedies in England demonstrates that health, disease, and desire could be provoked—and manipulated—by the air that one breathed. Though Shakespeare's Henry V imagines odoriferous hazards of the plague as a weapon of war, the continual presence of the disease in England's crowded urban centers made breathing there precarious as well. John Davies's account of the 1603 outbreak of the plague includes a similar image to Henry's: "London's Lanes . . . vomit their undigested dead, / Who by cart loads, are carried to the Graue."[27] Davies's metaphor ostensibly focuses on the cannibalistic and gustatory effects of the city consumed by plague, but the presence of cartloads of "undigested" dead on London's streets also offered an olfactory hazard that was both a cause and an effect of London's perpetual plagues.

Perfume was one way to combat foul, diseased air, particularly if released as a fume. Incense could fully penetrate a space as large as a cathedral with its smoky scent. Whereas the cost of incense, along with its religious symbolism, limited its use as a plague preventative, cheaper aromatics like rosemary were regularly burned in order to influence internal affective states and bodily health. Scented airs, dispersed in enclosed settings, attempted to regulate these environments. But their presence also raised profound questions about other dangerously unsavory, and unscented, contaminants that accompanied them.

Latent in the very definition of perfume is belief in the ability of smell to permeate bodies and environments; as such it offered a powerful way to combat unseen forces at work on both.[28] In his *Declaration of Egregious Popish Impostures* (1603), Samuel Harsnett reports on Catholic priests who subjected those suspected of heresy to forced fumigations with "brimstone, sulphur, Galbanus, S. Johns Wort, and Rue," and other "stinkinggere" like roses and rosemary: "Now I present unto your imaginations *Sara Williams* sitting bound in a chayre . . . with a pinte of this holy potion in her stomacke, working up into her head and out at her mouth, and her eyes, nose, mouth, and head stuffed full with the smoake of holy perfume, her face being sheld down over the fume till it was all over as black as a stocke."[29] Such uses underscore how the very act of breathing left one vulnerable to olfaction's penetrating powers. Though Harsnett describes these practices in order to critique them, he relies on a similar logic of olfaction to reveal not the stink of the devil but rather the stink of the Catholic church: "The Roamine Church, and her implements are of one and the same perfume, that doe out-smel the fuming lake spoken of in the *Apocolips*."[30] Whereas priests smell the devil in the scent of the young woman's body, Harsnett smells the apocalypse in their fumigations. Good smells can signal bad things; the stench of the apocalypse, for Harsnett, smells like a sweet perfume.

This association anticipates the conflation of sweet romance and apocalyptic death that characterized the use of rosemary during the plague. Such complicated metonymic associations between bodies and their scents demonstrate profound anxiety about the extent to which medicinal cures for the plague offered up their own olfactory hazards. Plague preventatives, medical texts, and theatrical representations of aphrodisiacs in the period all document a widespread belief in the influence of environmental stimuli on embodiment, particularly those stimuli that accompanied scented air. As one early-seventeenth-century commentator put it, "passion smelleth."[31] So much so, that Joyless, a character in Brome's *Antipodes* (1640), musing on the suspected "dishonour" of his wife, bemoans that he must be a "nose witness" to its "rankness," wallowing in the "bare suspicion of their filthiness."[32]

Environmental influences were key components of both Galenic and Paracelsian theories of health; extreme changes in climates, for example, could radically alter behavior, including sexual practices, gender norms, and even ethnicity.[33] Whether one followed a model of Galenic health, in which scents betrayed the body's inner balance of fluids, or Paracelsian models of disease and its cure, where extrinsic miasmas could attack the body and influence its health, scents marked the presence of other bodies, a fact that made them especially precarious when inhaled.

Early modern men and women were at a loss, however, to explain precisely which scents marked dangerous circumstances. Perfumes signified both healthy and diseased airs, creating dangerously intoxicating and infectious zones. The presence of sweet-scented airs could signify both the presence of the plague and its cure. Because the plague was a perpetual threat in early modern London, perfumed fumigations were a common smell. Scents used to combat the plague within quarantine spaces permeated other realms, reminding passers-by of the plague's unseen proximity. The smell of rosemary, in particular, symbolized the power of memory, memorials, and death in early modern literature and materialized the smell of shut-in households on early modern London's streets. Whereas more expensive aromatics could achieve the same function, rosemary's abundance and affordability made it a state-sponsored preventative for even the poorest neighborhoods. Likewise, pomanders, stuffed with expensive herbs and spices, enabled health practitioners and body-searchers to navigate dangerous realms.

Smells permeated the boundaries of the body. Perfumed scents, for example, were external to the body but were believed to alter its physiognomy.[34] Bodily odors influenced and indicated health. Perfumes worn on the body were believed to function as prophylactic barriers, defending the body from unseen contagions.[35] Yet emitting its scent also betrayed to others that one had recently traversed dangerous zones of contagion. Just as articles of clothing were thought to "have a life of their own," functioning as protective surfaces while "absorbing" other materials in the environment, so too did perfumes protect their wearers from (while announcing) the presence of bodily effluvia.[36] Smells thus did more than just influence health, they also marked one's social position; the clothes that one wore and the food that one ate shaped public identity.[37] Bodily odors—and the perfumes that sought to regulate them—were integral to understanding the relationship between early modern environments and the wide variety of bodies that inhabited them.

As many scholars of early modern medicine have noted, in hierarchal social systems, health was regulated through monitoring and calibrating humoral fluids.[38] Without such regulation, both one's bodily health and one's corresponding social position deteriorated, betrayed by bodily effluvia (like wasteful excrements of digested food or the skin's noxious toxins on clothing). Through what Kathleen Brown has termed "linen centered" models of cleanliness and what Michael Schoenfeldt has called the "great chain of eating," one might say that the body's odors were literally and figuratively woven into the social fabric of early modern culture.[39] Brown and Schoenfeldt emphasize that bodily smells were interpreted as an internal, fluctuating gauge of interior manifestations of health and as a reflection of one's social position. As Schoenfeldt concludes, "In the Renaissance, one could (in both senses) smell rank."[40]

Perfumes significantly complicated this osmology because scents affected such carefully calibrated systems. Although most contemporary scholars are silent on the issue of artificial scents and their impact on bodily health, early modern medical practitioners are not. Medical understandings of olfaction bridged Galenic humoral theory with Hippocratic notions of disease and contagion, and perfumes were crucial to maintaining a body's internal balance during exposure to contaminated air. Galenic humoralism postulated that diseases resulted from an endogenous, bodily imbalance; miasmic theories of contagion held that external polluted airs, vapors, and smoke could disrupt internal bodily humors.[41] In both theories of disease, the air that one breathed was central to understanding health. One needed to purge foul-smelling effluvia, but, once released, such ef-fluvia created profound environmental hazards. Whereas exhaling was always a healthy act, inhaling was not.

As I argued in the introduction, Helkiah Crooke's *Mikrokosmographia* offers insight into how perfume negotiated the paradox of olfaction and respiration. A middling sense, situated precariously between the perceptive senses of sight and hearing and the bodily senses of taste and touch, smell mediated between one's body and mind, one's animal and human instincts.[42] The nose, similarly positioned between the eyes and the mouth, represented olfaction's middling status within sensory hierarchies. But, as Crooke is careful to note, the nose merely drew air inside the body; only the brain could perceive odors.[43]

Although odors and vapors were invisible, they were not immaterial. Medical practitioners describe miasmas, odors, smells, and vapors as material objects circulating in the air. With every breath, one inhaled invisible particles that "touched" the brain."[44] Because of that immediacy, Crooke reasons that scents can alter humoral balances: "many odours of smels are able to refresh a man when he is ready to swound or faint away; they exhilarate or cheare the heart, and if we may beleeue Aristotle . . . the correct the distemper of the brayne."[45] The ability of odors to "refresh," "exhilarate" and "cheare" a body demonstrated that scents could radically influence both health and demeanor. In certain environ-ments, one might even "feed" on scents in the air. Crooke describes how the Eastern Indian *Astomy* survive on the scents of flowers and apples, how English cooks refuse meat after cooking it because they are too full from its "savours," and finally, how "trauellers" in the Low Countries were often driven into a "light madnesse" from feasting on the smell of "flowers and beanes" in the area.[46]

Rank odors, however, posed a serious threat: inhaling the wrong air could provoke disease, madness, even miscarriage. The "exhalations" from dead "car-kasses and muddy fens" infected the air and bred pestilence; a foul stench alerted medical practitioners to dangerous contagions in the air.[47] Similarly, smells from "ranke" and "impure" bodies, still very much alive, were so unpleas-ant they could "defile the spirits contayned in generative parts."[48] Greatest harm occurred when an external stench emitted a thin "spirit" or "vapor," "arousing" similar types of vapors already present in the womb.[49] Crooke concluded that wombs were "much affected with sauours and smelles; so that some haue beene knowne to miscarry upon the stench of a candle put out."[50] Although spontane-ous miscarriage was an extreme response, such an anecdote demonstrates how fraught with the dangers of smelling the act of breathing was for early modern men and women if external odors, vapors, and scents could permeate—and penetrate—brains, wombs, and generative parts merely through the act of inhaling. Since one could not control the environment and one could hardly stop breathing, this caused a certain amount of anxiety.[51] As one medical practitioner lamely concluded, it was best to breathe sweet smells: "The aire is one of the Elements whereof our bodies are composed, and without the inspiration, and respireation thereof we cannot liue and therefore it standeth much with our health that the aire which we receiue into our bodies, be sweet, wholesome, and uncorrupt."[52]

It was not always prudent, however, to rely on the environment to produce wholesome scents. As Crooke evocatively points out, odors and vapors are "wafted upon the wings of the wind or transported by the motion of the ayre."[53] A host of material circulated in the air, including bodily effluvia: "sometimes some seedes of the very substance of bodies that are of subtile parts are transported in the exhalation, which settling in the braine bring foorth fearefull accidents and strange effects."[54] Perfumes, however, could block foul scents from entering (and touching) the brain.[55] In describing the inability of the nose to perceive even the strongest scents of musk, civet, and ambergris without first inhaling them, Crooke describes a nose smeared and stuffed with scents: "for though you fill the Nose full with Muske or Ambergreese or other odoriferous bodyes; yea though you should annoynt the whole membrame with sweet oyles, yet you shall haue no perception of odours except you draw in the Ayre by inspiration."[56] Crooke, in using this image to demonstrate his point, draws on a common medical practice: the use of potent perfumes to protect the nose and to act as a physical barrier that would prevent other scents from entering the brain.

In certain environmental conditions, such precautions represented one's only defense against the plague. Early modern theories of olfaction accelerated plague-related fears of contagion.[57] One's own exhalations were as dangerous as another's. As Gail Paster argues, "All parts of the humoral body were capable of containing fumes and smoky 'fulinginous' vapors that could rise from the guts to the cranium, winds that roared and rumbled, sharp and vehement grippings, belchings, gross and clammy cruditites, fluids that putrefied and stank or, burning up, became 'adust,' seed that sent up poisoned vapors to the brain."[58] Regulating humoral flux produced a variety of smelly byproducts that engulfed and besieged the body. When located within discrete, indoor environments, like a church, shop, or household, these vapors and fumes created significant environmental hazards.

Early modern epidemiology understood the body's relationship to discrete, local environments to be a fundamental component of contagion. Hippocratic theories of "salubrity" defined health through air, soil, and water quality.[59] In the wake of England's devastating, but concentrated, plague outbreaks, early modern medical practitioners often resorted to Hippocratic, miasmic theories of contagion. Miasmas (derived from the Greek *miasma*, related to Greek *miainein*, "to pollute") were invisible, dangerous effluvia released from decaying organic matter, and thus associated with foul stenches.[60] Suturing ideas about airborne miasmas to Galenic humoral theory, medical practitioners labeled any bad smell, but especially bodily odors, as miasmas. Miasmas mapped neatly onto existing hierarchical understandings of "ranke" and "impure" bodies, and they provided a theory that could explain England's random patterns of plague contagion.

According to most accounts of England's sixteenth and seventeenth century plagues, communities monitored contagion. Patterns of infection were thus described in localized terms rather than nationwide geographies: homes, streets, and parishes. Plagues struck London parishes in 1578, 1593, 1603–10, 1625, 1636, and 1665, but their geographical scope and influence was erratic, striking some parishes repeatedly while skipping others completely.[61] With the exception of the plague of 1625, known for its extensive scope, fears about the nationwide spread of contagious air far exceeded actual patterns of contagion.[62] Certain geographical regions were believed to be high risk and were known for their dangerous, diseased vapors and fumes (as well as insidious rebellions), but, for the most part, contagion was thought to be concentrated in smaller pockets of larger communities.[63] The poor, the "masterless," the ungodly—and the areas of London in which they congregated—were specifically targeted as infectious.[64]

Confinement was integral to halting the spread of disease. Royal orders prioritized stopping contagion over administering cures and advised local officials to regulate contact with those identified as potential sources of infection: communities of vagabonds, family members residing in already-infected homes, and church officials that advised helping either of the above. London's liberties (and its cultural attractions) were viewed as particularly dangerous. In rural areas, markets represented a primary concern. Once infection occurred, social isolation was swift. Royal orders specifically warned preachers not to encourage parishioners to help those in need.[65] Plague wardens were instructed to nail, or "shut-in," the afflicted inside their homes, to mark the doors with a red cross, and to deliver necessary goods but never enter infected spaces.[66] Searchers, usually poor, elderly women in the parish, who were the only ones other than doctors and apothecaries to traverse these infected spaces, interpreted signs of infection and reported deaths.[67] Both wardens and searchers marked their exposure to the disease through visible cues, carrying white candles or wearing black and white lace caps that signified whether they had recently left a dangerous space, to announce that they may have absorbed infected airs in their clothing.[68]

For those afflicted, the plague resulted in a quick death. Most historians believe that mortality rates were between 60 and 80 percent, with death occurring most likely within the first week.[69] Confinement of an afflicted household, however, lasted no less than a month, trapping those not infected in dangerously stifled air.[70] Perfumes were believed to be potent curatives; royal orders published a host of remedies for perfumes to "correct the air."[71] The strong smell of decaying corpses lingered in contagious zones, underscoring associations between plague miasmas and bodily matter. Miasmas could travel long distances and thus required strong preventatives. Animal-based perfume ingredients, like musk and civet, could counter miasmas by weighing them down: "[N]oxious miasmes exhaling the cutaneous pres of the bodies of animals do infect the aer at a greate distance with malignants & subtle particles, without any difference in the weight of bodys, which do attest the great lightness of subtle effluvious which may be probed by many weyed instances signifiying when odoriferous bodys have emmitted a greate store of effluvias they do emmitte a small diminution of weight as it plainely appeares in muske and civet."[72] To use musk and civet to prevent miasmas from traveling, however, further amalgamated healthy and diseased air, conflating perfumes with dangerous contagions.

Wind of any sort reflected airborne conflicts between healthy, wholesome airs and contagions. Describing the effect of rain on plagued areas, one author notes that so many "antidotes" and plagued vapors are released into the air that they produce a stormy gale, "emulating the impetuous waves of water arising from violent gusts of high and boisterous windes when one part of aer is impestously protruded by another."[73] Falling on arsenic, cinnamon, mercury, and the like, rain counters "malignant" vapors, producing "great tumults" and violent motions in the air.[74] Air quality was of particular concern to those working in London's urban markets. Plague orders targeted urban locations like markets and advised monitoring of their public streets and waters: "for the uncleane keeping of the streetes, yeelding as it doth noisome and unsauory smells, is a meanes to increase the corruption of the aire, and giueth great strength unto the pestilence. Also, that all the ponds, pooles, and ditches about the Citty, if they yeeld any stincking and noysome smells, that they be scoured and cleansed, for there ariseth from them an euill and unwholesome aire, which furthereth the corruption of the aire and worse will do in hotter weather."[75] As this advice suggests, summer months were particularly dangerous.[76] Fusing unsavory smells with unwholesome air, medical authors attributed the marked increase in deaths during summer months to increased exposure to publicly shared air.[77]

To counter these spring-induced fevers, published plague preventatives advised lighting bonfires of oak trees, juniper branches, and perfume pastilles in public streets and in homes to "purge and alter a corrupt and unwholesome aire."[78] One author wished that such perfumes "were used through the whole Citty."[79] Although plague recipes often advised the very wealthy to purchase stronger, more expensive perfumes, cheaper fumes could be made by burning any strong scent, especially rosemary. Royal orders stipulated that town officials and merchants were to provide the necessary ingredients to those infected if they could not afford them.[80] Those in the household who were not afflicted were advised to prepare charcoal fires and aromatic "fumes" of either "great" or "small" expense (depending on what they could afford) at least once a week.[81] Recipes called for aromatic ingredients such as musk, civet, frankincense, juniper, rosemary, bay leaves, or, for those unable to afford such ingredients, herbs and onions.[82]

Continental approaches to the plague significantly differed from English methods. Authorities there preferred to round up the ill and transport them to hospitals rather than shut them in. But fears about dangerous airs led most medical practitioners across Western Europe to defend themselves from unseen contagions with thick perfumes and protective coverings. Italian, French, and German doctors covered their entire bodies in long cloaks with large beaks, or noses, stuffed with aromatic herbs and vinegars.[83] Such measures (and outfits) were rarely adopted in England. For those who traversed infected spaces, or for the merchants and urban poor who could not afford to abandon dangerous urban locations, navigating areas of infection required a host of new medical—and perfuming—technologies.[84] Royal orders recommended that a fume be prepared, but advised that the doctor should always stay closest to the fire, and never allow the infected body to cross betwixt him and the fire.[85]

These suggestions were easy enough to follow, but offered little surety to wary medical practitioners. Those traversing plague-ridden parishes needed handy protection for vulnerable bodily orifices. If one knew he or she was entering an infected area, plague preventatives advised "smothering" oneself (and one's clothes) with "some sweet perfume" as well as bringing along "some Nosegay, nodule, or Pomander . . . which you must alwaies smell unto."[86] Whereas medical practitioners in continental Europe protected themselves with heavy clothing, English men and women sought instead to create aromatic boundaries around themselves. To navigate plague-ridden households or parishes, the English relied on strong perfuming ingredients to create a distinct, environmental zone around their body.[87]

Smaller, portable perfume dispensers, known as pomanders (from the French *pomme d'ambre,* or apple of ambergris), were increasingly used throughout early modern Europe to protect the nose from surprise engagements with foul air.[88] The term is just one of many references that testify to the importance of ambergris in English culture, which I explore in chapter 5. Yet *pomander* also evolved beyond its original meaning of a scented ball of aromatic paste and came to signify both the elaborate metal containers and the scent ingredients they held.[89] For those without protective clothing or headgear, pomanders offered a ready defense against foul airs. So popular were these small dispensers that sixteenth-century prayer books often metaphorically invoked them as a spiritual defense. For example, Thomas Becon's 1563 *Pomander of Prayer* (published the same year as a disastrous outbreak of the plague that wiped out over 20 percent of London's population) metaphorically links his book of prayers to these earthly counterparts.[90]

Decoratively worn around the neck or attached to a girdle, pomanders were handy tools used to ward off potential infection. Most fit within the palm of a hand, with anywhere from four to eight small compartments for spiced ingredients. Others held a sponge that had been dipped in strongly scented waters. Some were shaped into symbolic motifs such as books, armor, ships, snails, fish, or skulls. Most pomanders did not indicate which types of scent ingredients were found inside them.[91] Many recipe books, however, document how these decorative objects were used to ward off infection, listing a host of recipes for the aromatic pastes or scented sponges to place inside.[92] Pomander recipes specifically marked as plague preventatives called for strong perfuming ingredients such as civet and musk, mixing them with zedoary (a known diuretic in the period that produced a spicy, hot, smell similar to cardamom, galingale, or ginger and attempted to purge the body of contagion).[93]

Wholesome scents from nature were preferred, but when these were unavailable, dried scent ingredients provided a good substitute in a pomander: "And I counsell al men that they auoyd all places of infection, all stinking and noysome smels, and when they are disposed to walke, that they walk in gardens, or sweet and pleasant fields, but neither early nor late at night, I haue set downe the making of a good pomander."[94] Although perfumes were useful preventatives, their material association with plagued air accrued more insidious qualities. The random, and concentrated, nature of early modern plague outbreaks led many to interpret contagion as the will of God; perfumes, even when used medicinally,

Pomander (Italian, circa mid-fourteenth century). Silver gilt. The shape as an "apple of love" and references to Paris's dilemma indicate its association with amorous gifting rituals. Its four hinged compartments, however, link it to plague prevention. Photo © Victoria and Albert Museum, London

could be seen to interfere with divine punishments. For some radical Puritans, perfumes were still conflated with the Catholic rituals involving incense; for others, with sins of luxury, vanity, and vice. Philip Stubbes, in his 1583 *Anatomie of Abuses,* concluded that "costly" pomanders and perfume ingredients could "denigrate, darken and obscure" one's "spirits and sences" as easily as they could prevent infection: "[P]alpable odors, fumes, vapours, smells of these musks, cyuets, pomanders, perfumes balmes & suche like ascending to the braine, do rather denigrate, darken and obscure ye spirit and sences, then either lighten them, or comfort them any manner of way. But howsoeuer it falleth out, sure I am, they are ensignes of pride, allurements to sinne and prouocations to vice."[95]

Similarly, Richard Braithwaite argued in his 1620 *Essaies upon the Five Senses* that "[s]ome are of opinion, that this peculiar *Sence* [of smell], is an occasion of more danger to the body than benefit, in that it receiues crude and vnholesome vapours, foggie and corrupt exhalations, being subiect to any infection; it is true."[96]

Pomander (Western Europe, circa mid-seventeenth century). The six hinged compartments, decorative floral engraving, and ring all suggest that it was made to be worn on the body, containing and protecting smells from escaping or entering it. Photo © Victoria and Albert Museum, London

Perfume, necessary to protect oneself from the plague, also marked moral culpability, making one vulnerable to other metaphoric and material dangers.

Perfume, Pleasure, and Contagion

In a wide variety of cultural materials, olfaction was metaphorically linked to the dangers of sensuality as much as it was materially linked to disease. Worried about the impact of "naughty" goods, Queen Elizabeth's chief minister William Cecil decried "an excess of silkes . . . wines and spyces" in London's markets, and that such an increase in goods would surely lead to an increase in robbery and debauchery.[97] Similarly, London, besieged by plagues, was commonly referred to as Sodom in popular pamphlets: "[M]any preasing into infected places and the leawdnesse of others with sores upon them perfuming into the open airee . . . and London is situated as pleasantly as Sodom and her sister Citty as before the Such over for the same sins we have committed."[98] Whereas sensuality is not always dangerous—or associated with perfume—perfume itself, as a tool of sexual incitement, contained the risk of disease.[99] To perfume the air with bad smells was potentially noxious, just as perfuming the air with good smells was salient (and potentially arousing).

Fifteenth-century allegorical representations of the sense of smell represent perfume as a tool of sensual pleasure rather than a preventative of sin and disease. Iodocus Badius Ascensius's allegorical woodcuts (1503) of an all-female "ship of fools" represent the sense of olfaction as two women gathering floral nosegays while a third sniffs a pomander bought from a perfume vendor. Aligning the sin of pride with perfumes used to combat pestilence, Ascensius's picture suggests that women's bodies emitted dangerous smells and therefore required cosmetic nosegays and medicinal pomanders. Similarly, Philip Stubbes states that the "sweete Pride" of women is a stench found in hell, noting that "dangerous" and "costly" perfumes create a plague of their own not only "where they be present" but also lingering in all of the places they have been for "a weeke, a moneth, or more after they be gon."[100] Stubbes's diatribe against women's perfumes (issued forth from women's "delicate" bodies) constituted a dangerous biohazard. Whereas Stubbes's puritanical fears about bodily scents are hardly representative, the lingering perfumes about which he writes invoke the logic of olfaction behind plague quarantines—that air quality was spatially and temporally difficult to control—and tapped into very real fears about the influence that scents could exert even without the presence of other bodies.

Proximity to other bodies made sweet scents like rosemary potent for reconfigurations of erotic love after plague outbreaks. Lyric celebrations of a beloved often included reference to odors, scents, and perfumes; in such examples, sniffing perfume is often linked to experiential, sexual knowledge of a loved one's body. Edmund Spenser, in Sonnet 64 of his *Amoretti* (1595), describes the erotic effect of proximity to his lover's body through her scents. Breathing is an erotic act: "Comming to kisse her lyps, (such grace I found) / me seemd I smelt a gardin of sweet flowers from them threw around."[101] Flowers, presumably disturbed by her breath, emphasize the temporality of the embrace in what otherwise represents a conventional blazon.[102] She smells of gillyflowers, roses "red," "bellamoures," "pincks but newly spread," "a Strawberry bed," "a bunch of Cullambynes," and "yong blossom'd Jessemynes."[103] Despite such a floral haze, Spenser observes that his lover's bodily scent is potent and distinctive: "Such fragrant flowres doe give most odorous smell, / but her sweet odour did them all excell."[104]

In countless literary examples like this one, smelling is imagined as a dangerously intimate erotic act. In his many poetic celebrations of his beloved, Julia, Robert Herrick compares perfume's odiferous pleasures to his lover's scent. In "Love Perfumes All Parts," Julia's body issues forth powerful aromatics of incense, ambers, and musk that culminate in an erotic coupling between Jove and Juno: "IF I kisse *Anthea's* brest, / There I smell the Phenix nest: / If her lip, the most sincere / Altar of Incense, I smell there. / Hands, and thighs, and legs, are all / Richly Aromaticall. / Goddesse *Isis* can't transfer / Musks and Ambers more from her: / Nor can *Juno* sweeter be, / When she lyes with *Jove,* then she."[105] Her lips smell of incense, but her hands, thighs, and legs are richly aromatic in their own way, more potent than musk or amber. In the "Pomander Bracelet," he savors Julia's "richly redolent" pomander beads, not for their perfumes but for the scent of "her that did perfume the Pomander."[106] Given the pomander's strong association with the plague, Herrick articulates a dangerous odiferous pleasure. Rather than perfume, he prefers her "richly aromaticall" bodily scents—scents that the pomander is conceivably meant to protect against. Perhaps noting the prevalence of erotic metaphors celebrating the pleasures of smelling Julia's body, the title plate of the 1648 edition depicts a bust of the poet with small angels hanging floral wreaths on his nose, perhaps trying to protect the poet from the dangers latent in sniffing other bodies, perfumed or otherwise.[107] Innumerable literary allusions from the period contain representations of women's bodies issuing forth sweetly scented vapors and scents that are imaginatively smelled by men.[108]

Olfaction's ability to refresh or corrupt the body had gendered implications for early modern medicinal prescriptions addressing sexual health as well as disease. Pomander bracelets, like Julia's in Herrick's poem, were one of the few pieces of moveable property bequeathed to daughters.[109] In her will, dated roughly 1624, Anne Clifford left her daughter a "bacelett of little pomander beads, sett in gold and enamelling, containing fifty-seven beads in number, which usually I ware under my stomacher."[110] Many reports note that, prior to her execution at Fotheringhay, Mary, Queen of Scots, "suffered her two women to disrobe her of her chain of pomander beads and all other her apparel most willingly, and with joy rather than sorrow, helped to make unready herself."[111] Disrobed and facing her executioners, her body is vulnerable in particularly gendered ways. Stripped of her clothing and pomander beads, she is made "unready," visually exposed to the crowd and its air.

The ability of perfume to penetrate the body had gendered connotations as well. Portraits from the period show that men wore pomanders around their

Robert Herrick, *Hesperides* (London, 1648). Title page. By permission of the Folger Shakespeare Library

necks, while women wore them hanging from their girdles.[112] The placement of the pomander close to a woman's womb was for two purposes, to protect her privy parts from external smells and to protect others from her bodily smells. This common sartorial practice suggests that early modern men and women believed that the womb, like the nose, was an organ of olfaction. Crooke explains that perfume inspires pleasure in the womb of any "crauing creature," but that it provokes particularly strong responses in women: "[The womb] will in a manner descend or arise vnto any sweete smell and from any thing that is noisome: which is the reason that many women are so easily offended with the smel of muske or other perfumes taken at the nose, for that the wombe moueth vpward vnto them; and in the fit of the rising of the mother, we apply burnt feathers and such like noysome vapours to the nost to driue the wombe downeward again, as also sweete and odoriferous suffumigations to draw it downeward to the owne naturall seate."[113] Crooke's observations suggest that both nose and genitals were believed to be organs of olfaction. Perfumes were medicines that returned the womb to its own "seate." Alluding to the practice of fumigating the womb to regulate the courses, Crooke aligns perfume with women's vaginal health as well as with their pleasure.

Ambrose Paré elaborates on this practice, further aligning perfume's medicinal applications with gynecological ones. In his *Workes,* published in many editions in the seventeenth century, Paré outlines how to cure both "strangulated" and "dropped" wombs by applying a perfumed pessary or aromatic "fumigations." To draw out evil "vapours," Paré used olfaction: "aromatick" scents were "received" into the womb and the fundament to correct a vaporous imbalance. Paré outlines a stunning litany of scents that pull on the womb from each of the body's orifices:

> There may be a fumigation of Spices to be received up into the wombe, . . . the matter and ingredients of sweet and aromatick fumigations, are Cinamon, Calam., Aromat.Lig.Aloes, Ladanum, Benzoin, Thyme, Pepper, Cloves, Lavender, Calamint, Mugwort, Penni-royal, Alepta moschat, Nutmegs, Musk, Amber, Squinant, and such like, which for their sweet smell and sympathy, allure or intice the womb downwards, by their heat consume and digest the thick vapours and putrefied ill juice. Contrawise, let the nostrils be perfumed with fetid and rank smells, and let there be made Gum. Galbanum, sapenum, ammoniacum, assafaetida, bitumen, oil of Jeat, snuff of a Tallow-candle when it is blown out, with the fumes of Birds Feathers, especially of Partrides or Woodcocks, of Mans hair, or Goats hair, of old Leather, of Hose-hoofs, and such like things burned.[114]

Paré's observation that sweet smells "allure or intice" the womb downward implicitly links wombs with the appeal of perfume. The womb, as a site of ill-scented humors, was particularly vulnerable to perfumes: perfume "intices" while "fetid" and "rank" smells repel it.[115] Scents also operated as diagnostic tools that could discern both uterine health and virginity. To test a maiden's virginity, Laurent Joubert suggests in his 1579 *Erreurs Populaires* that the young woman should smell "some broken patience dock leaves"; if she did not instantly produce urine, then she was not a virgin.[116]

Paré offers a more explicit link between perfume and sexual practices: if a married woman suffers from strangulation of her courses, he recommends that her entire belly be smeared in an aromatic paste (using similar ingredients to those listed above), and "if she be married, let her forthwith use copulation, and be strongly encountered by her husband."[117] For unmarried women, Paré prescribes another perfumed remedy: "Let the Midwife anoint her fingers with oleum nardinum, or moschetalinum, or of Cloves, or else of Spike mixed with Musk, Ambergreece, Civet, and other sweet powders, and with these let her rub and tickle the top of the neck of the womb which toucheth the inner orifice . . . for so at length the venomous matter contained in the womb, shall be dissolved and flow out, and the malign, sharp, and flatulent vapours, whereby the womb is driven as it were into a fury or rage, shall be dissolved and dissipated."[118]

As Paré suggests, orgasms were tools used to "purge" the body of toxic humors; "sexual incontinence" was a vital component of regulating health.[119] These excerpts indicate, however, that perfumes or odors were linked to the erotic action of "tickling" and that such tactile and olfactory pleasures were connected. Such simulated "purges" were equally viable alternatives to orgasms (and often were combined with them), they and provide insight into why diuretics were linked to aphrodisiacs in the period.[120] Perfumed pessaries were similar in both form and function to sexual aids; long, cylindrical tubes, with small perforations, they were designed to be strapped into "the neck of the Womb."[121] These devices, administered by a midwife, were designed to inspire orgasms in order to purge the body of foul humors. Perfumes were necessary to counter such foul exhalations. The visual verisimilitude between pessaries and dildos suggest that perfume may have played a larger role in the history of sexuality than we suspect.

Perfumed pessaries functioned as both medicinal and sexual aids.[122] Numerous jokes in early modern popular culture link perfumed glass "dildos," similar in shape to pessaries, with women's sexual pleasure. Ben Jonson, in *Cynthia's Revels,* jokes about "perfumed dogs, monkies, sparrows, dildoes, and paraquettoes," connecting puns about perfume, sexually promiscuous women, and the intimate trifles found in fashionable women's bedchambers (and actions imagined to occur with them).[123] John Marston, in his *Scourge of Villanie,* also cites women's use of perfumed glass vials as sexual aids, joking about the scent of monkeys, instruments, and other "female" luxuries.[124] Marston furthers the link between perfumed "pets," dildos, and women's genitalia, punning on the aromatic power latent in a filthy spaniel's scent: "My spaniel's paunch, who straight perfumes the room / With his tail's filthy." Spaniels, the quintessential seventeenth-century English lapdog, were known for their keen sense of smell and their "metonymic association with women, wealth, and outlandishness."[125] This spaniel's scent, however, permeates the room, linking his stinky tail with the power of women's laps to alter an environment.

Women's laps were potent sources for misogynistic jokes about scents and sexuality in early modern England. Although it is tempting to interpret such humor as yet another example of a seemingly timeless misogynist riff on women's bodily scents, these jokes evince early modern cultural anxiety about the potency of body odor and rank air. Marston's character had reason to fear the spaniel's scent: read within the context of plague preventatives, women's bodily smells were agents that could alter environments. Although medical treatises focused solely on the ability of scents to alter women's bodies, other cultural materials documented fears about women's ability to manipulate smells in order to affect other bodies. Women's aromas were themselves rumored to be strong aphrodisiacs; women who hoped to capture their lovers' hearts were advised to give gifts of "love-apples," small apples that had been stuffed into their armpits until they were soaked with sweat.[126] In Marston's *Malcontent* (1604), Prepasso, a gentleman usher, jokes about the "ill sent" of a brothel house and calls for perfume to protect himself: "You thinke you are in a brothell house, do you not? This roome is ill sented. [*Enter one with a perfume.*] / So, perfume, perfume; some vpon me I pray thee."[127] Marston's joke, of course, is that perfumes were also aphrodisiacs in the period.[128] By the end of the seventeenth century, with the dissipation of outbreaks of plague, pomanders were known more for amorous, rather than medicinal, applications: "Pomanders or perfumed bracelets may be used, and by their odiferous scent conduce much, Ladies, to the making your Captives numerous, though they bind only your Arms, yet may they take Men your Prisoners."[129] Although such a recipe did not literally suggest ways to "capture" male prisoners, its playful link of olfaction, binding, and captivity gestures to the complicated history of pomanders as tools for negotiating shut-in spaces.

Staging Perfume

As such literary sources suggest, sexual arousal, like health, was a precarious bodily condition, easily manipulated through olfactory stimuli. The scent of rosemary, along with other sweet perfumes, marked debauched zones of peril and pleasure; the ability of pleasant odors to "refresh," "exhilarate," and "cheare" a body (along with their ability to tire, weaken, and harm it) made perfume both a tool that fought disease and an aphrodisiac that could inspire contagion. Representations of scented air on stage, in particular, document ways perfume was used to manipulate both health and sexuality. Dramas like Middleton's *Women Beware Women* (1623) and Shakespeare's *Midsummer Night's Dream* represent perfume as a sexual stimulant and noses as vulnerable orifices. As these plays document, the efficacy of scented airs depended both on the environments in which they were deployed and on those who were in control of the perfumes. *Women Beware Women*, in particular, draws explicit connections between perfumes, urban experiences of plague, and dangerous sexual proclivities. Though set in Italy, the play's smellscape builds on Londoners' intimate experiences with disease and city life.

Perfume, confinement, and illicit desire are at the crux of the play. As its titular warning indicates, women are equal participants in enacting the play's pharmacological fantasy of misogyny, incest, and rape.[130] Many critics have noted that the play stages social traffic in women: as Isabella dismally concludes, "Men buy their slaves, but women buy their masters" (1.2.174).[131] Marriage is an economic negotiation, and sexual favors are bought, bargained for, or taken with or without women's consent. Set within a few wealthy households, the play examines Florentine cultural and material environments in domestic settings. The action begins with a serious social transgression: lowborn Leantio, recently married to the beautiful (and wealthy) Bianca, brings his Venetian bride home to Florence. The Duke of Florence spots Bianca at a civic parade with her mother-in-law and desires her for himself. Livia, a wealthy Florence aristocrat, invites Leantio's mother and Bianca to her home and enables the Duke to rape Bianca while she plays a game of chess with Leantio's mother. When Leantio returns to find his wife has become the Duke's "mistress," Livia seduces him. Meanwhile, Livia helps her brother, Hippolito, broker the marriage of his niece—and lover—Isabella to Ward, a rich, and witless, young heir so he can continue his illicit, incestuous affair. Desire and death fuse in the play's final act, as poisonous incense, arrows, and potions transform Isabella and Ward's marriage masque into a grisly death dance.

In this troubling drama, women are players and pawns in social alliances that result in sexual violence. As Nathaniel Richard's commentary verse in the play's opening pages surmises, Middleton's play, rife with disease, pornography, rape, and poison, brings "hell-bred, malice, [and] strife" to life on stage: "he knew the rage, madness of women crossed; and for the stage fitted their humors."[132] Given the its many jokes about the effects of "sorcery, drugs, and love-powders," *Women Beware Women* suggests that women are capable of manipulating their own bodily humors—and those of others. Livia advises Isabella—soon to be married to the inane Ward—that a woman's sexual liberty is a floral substance: "You have liberty enough in your own will; you cannot be enforced: there grows the flower"

(2.1.115–17). Ward extends Livia's floral metaphor. He puns that Guardinio, his uncle, seeks to "garden" for a wife: "My guardiner has no more wit than a herb-woman, that sells all her sweet herbs and nose-gays, and keeps a stinking breath her own pottage" (2.2.90). Both Livia and Ward imagine sexual desire—whether virtuous and virginal or stinking and diseased—as perfume. The play quickly materializes these literary allusions as Bianca is lured into the erotic art gallery. The Duke enters unseen, like a sinister "vapour that when the sun appears is seen no more," and ravishes her (2.2.315).

The Duke's sexual advances seem to offer an olfactory representation of rape and ravishment, materializing in the play as a dangerous fog of smoky air through which he attacks.[133] In the play's third act, just prior to Leantio's realization that he has been cuckolded, he describes the concordance between love and marriage as olfactory bliss: "The treasures of the deep are not so precious / As are the concealed comforts of a man, / Locked in a woman's love. I scent the air / of Blessings when I come but near the house / What a delicious breath sends forth" (3.1.81–92). "Locked" in a woman's love, Leantio imagines wedded bliss as a sweetly scented house, fragrant with blessings.

As the third act progresses, the household air becomes stagnant and poisoned, linking the play's staging of desire to experiential memories of the plague and shut-in households. Leantio's metaphoric invocation of "fragrant blessings" reveals his delusions about his marriage. Bianca smells, in comparison, a changed presence in the house, remarking on its artificial "strangeness." Declaring that it is full of "defects as ever gentlewoman / made shift withal to pass away her love in," Bianca comments on the prevalence of cushion-cloths, lacy "cut-works," and "silver-gilt casting bottles" hung in her bedchamber, linking the prevalence of luxurious goods like liquid perfumes to defects in her bedroom. Even the dullard, Ward, notices the changed atmosphere. When Isabella woos him with a song praising beauty (but verbally comprised of "compounds" and "physic-work"), Ward snidely remarks that Isabella's beauty is cosmetic. Isabella's body issues forth delicious scents, but Ward says that his pleasure in kissing her could result solely from her perfumes. His gross sexual advances reveal that he is aroused by her perfumes but recognizes that they are part of an artificial construction of beauty and desire. Before consenting to marriage, he surveys her body, asking, "is that hair your own," before "peering" at her teeth, walk, and "up her skirt" (3.3.54–63). Noting her "most delicious scent," Isabella's perfumed breath arouses his pleasure, yet it inspires synesthetic confusion: "Methinks it tasted as if a man had stepped into a comfit-maker's shop to let a cart go by, all the while I kiss'd her." Rather than celebrate his lover's bodily scent, Ward conflates it with a distinct urban space: a spice shop. Her body is imagined as a distinctly urban space, one that he's wary of entering.

Ward's paranoia materializes with the discovery of Livia's plans, represented on stage as perfumed letters. Such letters—as stage scents—are reconfigured as a plague: when Leantio confronts Bianca (and threatens a "plague" on the house), he leaves Livia's "perfumed" love letters behind. Bianca, reading and smelling Livia's scented signature, realizes that she has been duped: "Why, here's a bawd plagu'd home" (4.1.75). Her response, after he leaves, is to purge the air of pestilence: "I'll have this sauciness / Soon banish'd from these lodgings and the rooms / Perfum'd well after the corrupt air it leaves. / His breath has made me almost sick, in troth" (4.1.104–10). Misinterpreting her nausea as the result of Leantio's rank breath, rather than Livia's perfume, Bianca only adds more scents to the already clogged household. As Bianca prepares to marry the Duke, the Cardinal condemns the marriage, declaring that "every sin" is a flame spreading throughout the region, fed on a "big wind of popular breath" (4.1.206–10). Perfume is no longer a liquid, but rather an airborne fog, as the troubled household chokes on itself.

Such a point is made literal in the play's final act. Just as rosemary signaled both betrothals and funeral practices, Middleton's masque-within-a-play transforms the "mists" and "perfumes" of love into fumes associated with plagues, disease, and death. Isabella's scented smokes circulate throughout the entire house. The masque, which celebrates Bianca's marriage to the Duke, becomes a dance of death. Entering in floral garlands, lighting candles and censers, Isabella releases powerful fumes that envelop Livia: " 'I offer to thy powerful deity / this precious incense; may it ascend peacefully—' [The incense sends up a poisoned smoke] [Aside] 'And if it keeps true touch, my good aunt Juno, 'Twill try your immortality ere't be long" (5.2.72, 105–10). Such fumes work quickly: the "savour" overcomes Livia, who rapidly concludes, "this fume is deadly. Oh 't has poisoned me" (5.2.115, 130–31). Violent revenge engulfs the performance as theatrical props commit material violence: Livia, in retaliation, throws flaming gold on Isabella before her collapse; Guardinio falls through a trap door; and Hippolito dies by a poisoned cupid's arrow, all of which is seemingly set in motion by the change in atmosphere. Finally, the Duke, who earlier enacted a deadly—but metaphoric—act of "vaporous" violence, is mistakenly killed by Bianca's poisoned cup, which was meant for his brother the Cardinal. Bianca kisses the dead Duke, attempting to share his poisoned breath: "Give me thy last breath, thou infected bosom, / And wrap two spirits in one poison'd vapour" (5.2.195–96). The play's horrific conclusion comments on the disastrous effects of domestic perfumes.

Linking perfume's metaphorical ability to provoke sexual arousal with its material applications as a plague preventative, Middleton's play, fitted for "women's humors," anxiously explores the effect of breathing perfume within confined spaces of the home.

Staged in enclosed spaces like the Duke of Florence's house, perfume is imagined as a powerful aphrodisiac that defines spatial zones. As such, it offers a powerful counternarrative to perhaps the most famous use of floral aphrodisiacs in the period: Shakespeare's *A Midsummer Night's Dream*. Many scholars have analyzed Oberon's infamous use of a "little Western flower" to drug Titania. Given the play's obsession with visuality, Oberon's drug cannot easily be classified as a perfume. Yet its foregrounding of environmentally situated desire makes it a powerful counterpoint to Middleton's play. Set in the forest's fairy realm—a realm whose material counterpoint undoubtedly resonated with descriptions of new world forests (described in chapter 3) and the ways vision was thwarted within them—sexuality is more a tool of humiliation than of pleasure; characters rarely find sexual satisfaction in the forest.[134] The few who do are humiliated by such unions. The psychic and material interchangeability of desire in the forest culminates, as Jean Howard argues, in Titania's "humiliating" union with an ass.[135]

As Howard makes clear, Oberon stages Titania's sexual union as punishment. In honor of her friendship with an Indian votress killed in childbirth, Titania refuses to relinquish the Votress's son to Oberon. Frustrated in his struggles with his former queen over possession of the Indian boy, Oberon orders Puck to deliver him "a little western flower—Before milk-white, now purple with love's wound,—and maidens call it love-in-idleness" (2.1.59–61).[136] His goal is to smear Titania with it and exert power through sexual humiliation. Once drugged, she will surrender to his will since "the juice of it on sleeping eyelids laid / Will make or man or woman madly dote / Upon the next live creature that it sees" (2.1.69–72). Oberon's tools betray his plan: the power latent in the "little western flower" lies in its ability to arouse indifferently regarding sexual objects; this neutrality enables the expression of alternative desires. Wounded by Cupid's arrow, spotted by Oberon, and named by maidens, love-in-idleness contains powerful pharmacological properties that provoke instant arousal. Known in the early modern period as heartsease, and today as pansies, Oberon's flower, as Gail Paster argues, represents "a deliberate intervention in [Titania's] cognitive and affective processes."[137] Its love juice streaking her eyes, Titania is made "full" of "hateful fantasies" (2.1.69–72).

Such a pharmacological vision is striking in the ways olfaction is *not* implicated within it. Fantasies of bodily control focus on how an almost scentless flower can magically transform into a powerful philter. In her study of relationships between early modern drugs and theatricality, Tanya Pollard argues that pharmacology is at the heart of the exploration of love in *A Midsummer Night's Dream*. Titania's drugged, amorous visions reflect a "troubling" metaphor about the effects of "strange and exotic" drugs within early modern culture, particularly within the environmental realm of the theater.[138] In Pollard's analysis, amorous drugs, as material properties or as theatrical metaphors, reflect increased awareness of pharmacological constructions of both health and desire. Pollard notes, however, that the play's "love-in-idleness" inspires tenuous effects: "Like Cupid, Oberon brings about a transformation, but not exactly—or only—the one he had in mind."[139] Whereas Pollard highlights the philter's application on Titania's eyes, linking the visions it inspires to those of the theater, I emphasize how the play nervously cites the olfactory realm of the theater and the efficacious role of smell within it by interpreting the play alongside Livia's perfumes.

To do so, I focus not on the smells of the forest, or even of the aromatic setting in which Titania is drugged (it is, after all, a floral bank where wild thyme "blows," violets "grow," quite "overcanopied" with "luscious" woodbine, "sweet" musk rose and eglantine), but rather on how olfaction shapes the rude mechanicals' presence within the forest realm (2.1.249–56). Rehearsing scenes for their play in the forest, Bottom (as Pyramus) and Quince emphasize an olfactory discord that prefigures the relationship between perfumed plagues and aphrodisiacs. Parodying generic lyric connections between a beloved's breath and the sweet scent of flowers, Bottom's malapropism reveals the material connections between beauty and perfumes:

> *Pyramus:* Thisby, the flowers of odious savours sweet,
> *Quince:* Odours, odours.
> *Pyramus:* Odours savours sweet:
> So hath thy breath, my dearest Thisby dear.
> But hark, a voice! stay thou but here awhile,
> And by and by I will to thee appear (3.1.70–75).

Thisby's breath is materially odious but metaphorically odoriferous. Bottom and Quince's emphasis on Thisby's olfactory transformation mimics the magical one about to occur: rather than appear to Thisby, Bottom appears to Titania.

Although critics have noted both Oberon's pharmacological maneuvers and their role in violently manipulating the sexual agency of Titania and the Athenians, few have explored the play's environmental staging of them. Consummated on an imagined bank of wildflowers, Titania's desire for Bottom, like Bianca's for the duke, suggests that airborne scents were devices that could potentially provoke—and usurp—sexual agency. *A Midsummer Night's Dream* offers a dark, but ostensibly comedic, metaphoric exploration of perfume's aphrodisiacal effects. It offers its own commentary on how sweet savors can become odious. Staged within the forest's wholesome, refreshing night air, and placed under male control, the pansy's juice would be overwhelmed by the scents of other flowers, staging a confrontation between the eye and the nose as the most powerful organ of desire. Interpreted in larger contexts of early modern contagions and aphrodisiacs, however, the play's fantasy of "love" in idleness is constructed with materials familiar to early modern audiences: perfumes. A host of materials circulated in early modern air, including contagions and aphrodisiacs. Controlling perfume meant controlling the health and sexual agency of oneself, as well as that of others in the household.

To mark its 2005 production of *A Midsummer Night's Dream*, the Royal Shakespeare Company commissioned a material and commercial counterpart to Oberon's drug: love-in-idleness, the perfume. Finding the smell of pansies alone to be too weak to capture the power of the love potion, Dr. Charles Sell, fellow of the Royal Society of Chemists, blended their scent with those of other plants found on Titania's riverbank. When asked by BBC News Four about the amorous effects of his perfume, or how best to apply it (whether to smell it or drop it on sleeping victims), the chemist jokingly warned the audience not to "try this at home." Implicitly citing both the history of violence latent in his perfume and jokingly deriding its efficacy as an aphrodisiac, Dr. Sell's perfume reminds us of the ways olfaction might bear a trace of collective memories from the past, namely the experience of perfume within the household. Titania's riverbank is rife with olfactory pleasure, yet Oberon's usurpation of it transforms the fairy queen's environment from one of pleasure to one of violence. Read against early modern discourses of disease, such scents render desire as an embodied, environmental symptom, the result of invisible, exogenous agents carried by the wind, rather than as a reflection of identity, sexual choice, or agency. Early modern English men and women, however, understood disease and desire as interrelated affective states: both could circulate in the air as external agents that worked on the body, a fact made a bit too perceptible by the presence of perfumes.[140]

CHAPTER FIVE

Oiled in Ambergris

Ambergris, Gloves, London's Luxury Markets

In 1633, after a decade of success divining for London society about the fertility of Queen Henrietta Maria, Lady Eleanor Davies was fined and imprisoned for comparing King Charles I to the biblical tyrant Belshazzar, the last king of Babylon, and prophesying his death.[1] Almost a year later, while still in prison for treason, Lady Eleanor Davies's spirits were fortified by an angelic visitor, who, for the "space of an hour," held "throne" on her bed and "sent thither the Holy Ghost."[2] As the angel prepared to leave, he blessed her with his right hand. His blessing offered her spiritual sustenance and material proof that she had been touched by an angel who left "such an odoriferous scent when he was gone."[3] In letters to her sisters and in later publications, she described the angel's heavenly scent as emanating not from his divine body but from a rather earthly accessory: his right glove, which was "all oyled with Ambergreece, the spirit thereof proceeding from the Leather, so far beyond expression, as if it were invisible food."[4]

Lady Davies's career as a prophet was long and varied; this was not the first time she claimed to have been visited by a messenger of God, nor was it the last time she was imprisoned for treason. Later episodes led to her condemnation as a lunatic in 1637 and her confinement in Bedlam, but for modern historians it is this striking episode with the fashionably gloved, pleasantly scented angel that defines her career as a seventeenth-century prophet; most conclude that it proves she was mad.[5] For Lady Davies's contemporary audience, however, the angel's divine spirit, "proceeding" from leather gloves, had another ephemeral antecedent: the widely recognizable, commercial scent of ambergris.

Davies's vision occurred in 1633, at the end of what one fashion historian describes as a fifty-year "mania" in England for perfumed gloves—one year before the incorporation of the Worshipful Society of Glovers, fourteen years after the incorporation of the Worshipful Society of Apothecarists, and well after the commercial success of the New Exchange.[6] The scope of this trend and its subsequent impact on the economies that produced luxury perfumed gloves transformed the commercial perfuming industry in London; at its start, perfume was a verb; by its end, a noun. Read against such economic histories, the angel's perfumed glove signifies more than Lady Davies's eccentricity, madness, or perhaps even her talent for religious allegory. Like the smell of damask rose gardens on Bankside and the burning scent of rosemary issuing from its quarantined households, the scent of ambergris defined London's luxury markets. Together, they suggest that London's urban spaces were defined by their smells.

In this chapter, I explore the contours of this landscape, particularly its luxury markets, by examining the angel's earthly, odiferous accessory: his glove, perfumed in the "Spanish" style and heavily scented with ambergris. In late sixteenth-century England, a wide variety of guild-members, artisans, housewives, and merchants were "all oiled in ambergris"; many sought to control the lucrative production and consumption of these scented gloves. More than five guilds claimed ambergris under the sole purview of their organization; luxury markets like the Royal and New Exchanges enabled those shut out of the guild economies—namely alien men and women and native women—to become merchants of luxury commodities like perfumed gloves. At the same time, published and manuscript cookbook recipes document that perfuming gloves was also a common domestic practice performed by English housewives. Sharp distinctions between medicinal, culinary, pharmacological, fashionable, and cosmetic uses of aromatic ingredients like ambergris thus emerged as each group struggled to control the lucrative profit from perfume ingredients like ambergris. By the end of the 1630s, perfumer was a recognizable professional identity and perfumes, recognizable commodities.

Renaissance Gloves, Ambergris, and the Spanish Style

Lady Davies's angel is just one example of many that demonstrate the pervasiveness of perfumed gloves in early modern culture: letters, recipes, portraits, and guild records suggest that this trend affected almost every level of urban English life, including both imported spice markets and domestic leather industries. Renaissance gloves varied widely in style, shape, and color, and they served many utilitarian functions, including protecting laborer's hands from extreme weather and harsh activities. Even embroidered samplers from the period represent figures, in appliqué, wearing tiny leather gloves.[7] Gloves are prominently featured in Renaissance portraiture, serving to demarcate a subject's social status. Albrecht Dürer's *Self Portrait with Gloves* (1498), for example, elevates the painter's class status through fashionable attire.[8] Powerful royal or aristocratic figures in early modern portraiture often hold a single glove in their hands, symbolizing political and economic power. Wearing one glove while holding another indicated that the individual was about to enter the public sphere.[9] A cast-off glove symbolized the wearer's casual approach to expensive luxury goods, as well as the ability of such goods to circulate throughout early modern English society.[10]

Visual details in portraiture provide important clues to how Renaissance gloves were worn, exchanged, and received. Gloves were given as gifts; exchanged as tokens of betrothal, of desire and affection, of aggression; and worn as status symbols.[11] As love tokens, gloves worked especially well. A popular aphorism that adorned such gifts explains: "I send to you a pair of Gloves / If you love me, Leave out the G. / And make a pair of Loves."[12] Gloves were also gifted as wedding tokens, signifying that courtship had ended. The narrator of Robert Herrick's "To Rosemaries and Bays" explains that "MY wooing's ended: now my wedding's neere; / When Gloves are giving, *Guilded be you there.*"[13] Perfumed gloves were particularly valued in such rituals: scents, along with elaborate embroidery, denoted luxury and style. Richard Sackville, Earl of Dorset, was known for his collection of luxury gloves—an embroidered and scented pair for every outfit he owned.[14] And, as one anxious mother of a bridegroom wrote in 1611, "I could not get so many women's Jessamy gloves as [I] wrote for; and at the last was fained to pick upon cordinant for men and perfumed kid for women. I had them perfumed better than ordinary that they might give consent."[15]

Sentiments like this reveal that, by the end of the sixteenth century, perfume was an important indicator of the value of gloves. Perfumed gloves were much more expensive than those left untreated.[16] The costliest were in the "Spanish" style and scented with ambergris, though Italian and French gloves were also known for their elaborate perfumes.[17] For example, John Florio thought perfumed gloves important enough to include numerous conversations on how to procure them in his 1578 Italian phrasebook. Florio provides essential questions and phrases for English tourists hoping to purchase gloves from Italian merchants: "These Gloves, are they wel perfumed," "Who hath perfumed them," "Shewe me a payre of gloues," "I will buye a payre of Gloues," "And so will I too," "wil yea haue the perfumed or no," "I will haue them perfumed," and "Behold here is a good payre."[18] Reading the phrasebook, literary critic Jonathan Hope wonders about Florio's preoccupation, musing that "the subject of buying gloves, and the vexed question of whether or not they should be perfumed, comes up more frequently than you would expect."[19] Florio's translation of Montaigne's essays, for example, includes a segment on how the essayist's "thick" mustache held the scent of his gloves: "Let me but approach my gloves . . . to them, [and] their smell will stick upon them a whole day."[20]

Florio was not alone in his preoccupation with scented gloves. Countless plays in the period describe both the appearance and smell of gloves, suggesting that scented gloves were central to understanding a variety of social exchanges: Shakespeare's *Much Ado About Nothing, The Merry Wives of Windsor,* and *The Winter's Tale* all mention perfumed gloves; in *Macbeth,* Lady Macbeth infamously decries her bloody hands as perfumed; gloves are exchanged in *Troilus and Cressida* as an expression of love and as a symbol of betrayal, and in *Henry V* as a token of hostility and aggression; Deflores creepily penetrates the leather sockets of Beatrice's lost glove in Thomas Middleton and William Rowley's *The Changeling;* Bilbo, in Thomas Dekker's *Match Me in London,* a play set in a fictional Spanish court, describes his "aromatical, most odiferous" glove; Sir Glorious Tipto, in Ben Jonson's *The New Inn,* is careful to note that although his clothes are Italian, his gloves "are natives of Madrid" (2.5.57–69); and, in Jonson's *The Alchemist,* Sir Epicure Mammon boasts of his gloves "of fishes' and birds' skins, perfumed / With gums of Paradise and Eastern air" (2.1.40–46) and locates "Spanish titillation in a glove / [of] The best perfume" (4.4.13–14).[21] Mammon's dramatic boasts emphasize that the very best perfumes were those that "titillated" in the Spanish style.

London's early modern markets were piquant and heady realms, best navigated through a merchant's—and, implicitly, a consumer's—sensory experience of them. In the widely read *Merchant's Map of Commerce* (1634), Lewes Roberts, a London merchant, evocatively organizes England's commercial map through the art of "merchandizing." Merchandizing, Roberts argues, demonstrates a purveyor's general expertise in the mystery, skill, and art of trade, as well as in specialized knowledge of particular commodities; a merchant's senses are vital to both skills, linking commodities with particular bodily organs. Whereas "the eye" is a purveyor's most useful tool in achieving general expertise, the other senses are central to particular trades: "Some [merchants] are noted againe to require the *sence* of *feeling* to be assistfull to the *eye,* as where the *hand* is of necessity to be imployed, as is seen in *cloth* and such *commodities.* Some require the *sence* of *hearing,* as where the *eare* giveth a help to the *eye,* as is seen in some *mettalls, mineralls* and such like: and some againe require the *sence* of *smelling,* as where the *nose* helpeth the *eye,* as is seen in some *drugges, perfumes* and the like; and lastly, some requireth the *sence* of *tasting,* as where the *palate* giveth the helpe, as is seen in *spices, wines, oyles,* and many such *commodities.*"[22]

The five senses provide a legend for navigating Roberts's map of early modern London's markets. While Roberts marks olfaction as key to selling "certain drugs, perfumes, and the like" and taste as central to spice, wine, and oils, it is almost impossible for him to draw a sharp line between drugs, perfumes, and spices. That olfaction is linked to both perfume and drugs indicates its continued role as a medicinal cure. By 1635, however, perfume was also a valuable luxury commodity. Geographical schemas like Roberts's demonstrate that the senses were a key part of attempts to reorganize and control London's luxury markets, often defining (and redefining) economic and political relationships. And over fifty years' worth of guild struggles to define aromatics as either drugs, perfumes, or spices document that merchandizing perfume involved highly contested territory. As the economic history charted in this chapter reveals, the production, merchandizing, and consumption of scented gloves was a complicated, dangerous, and smelly affair, reconfiguring critical understanding of what it meant to traverse such market spaces.[23] To understand a map like Roberts's, one must account for the primary factor that defined these luxury gloves: the smell of their perfume.

According to a nineteenth-century chronicler of the glovers' guild in London, it took three nations to produce one pair of quality early modern gloves: Spain to dress the leather, France to cut it, and England to sew it.[24] Good gloves were more than the sum of their national or political parts; the very best gloves were produced using a variety of materials, knowledge, and labor that cut across national and political boundaries. Gloves were thus an amalgamation of quality leather, labor, and perfume dressings. By the end of the fad for perfumed gloves, several purveyors claimed to specialize in such luxury production, including glovers, grocers, apothecarists, milliners, perfumers, and haberdashers. And the aromatics used to dress them, including the most important ingredient— ambergris—were claimed under the purview of at least five (and as many as seven) London guilds.

For early modern men and women, the precise origin of ambergris, as Herman Melville famously wrote, was a "problem to be learned."[25] A waxy, odiferous, grey product of the adult male sperm whale, ambergris had no national provenance. Rather, it was "sea-born," as Andrew Marvell's seventeenth-century poetic description of it declared.[26] It can be found floating on the ocean or washed up, in hardened form, on beaches. Known to Persian and Muslim sailors and traders as "anbar" since the ninth century CE, ambergris was first encountered by European sailors during Spanish and Portuguese trading expeditions off the coasts of Africa, the East Indies, China, Japan, Australia, and New Zealand in the fifteenth century.[27] Almost immediately, it was valued for its strong and distinct scent.

Though we now know that ambergris is a growth that develops on the stomach and intestines of the whale, induced by undigested pieces of cuttlefish, which the whale defecates before it grows too large to pass, early modern men and women believed it to be either the "spawne" of the great whale or the "excrement" of a great sea fish.[28] Its greasy, grey appearance linked it to musk and other heavy, animal-based perfumes from sex glands, whose strong scent could counter the noxious smell of untreated leather. In the twentieth century, ambergris was usually macerated in alcohol and steeped for at least three years before it was used in perfume. Its smell has been described as salty, woodsy, and musky; because sperm whales have become endangered, it is rarely ever encountered in modern smellscapes, though its influence remains: synthetic ambergris is a common ingredient in modern perfumes. Though its French name, ambergris or "grey amber," emphasizes its visual resemblance to amber resin when in hardened form, its anglicized name alludes to its commercial applications, translating the French "gris" into the homophone "grease": in waxy, granular form, the greasy substance could be smeared onto leather objects.

The discovery of ambergris transformed Spanish perfumes, particularly those made in the south. The Muslim legacy of Al-Andalus—and the global heterogeneity that defined the Iberian Peninsula in the fourteenth and fifteenth century—influenced Spanish perfumers, whose scents primarily used musk, camphor, aloe, ambergris, and saffron (rather than sandalwood, spikenard, frankincense, myrrh, and rosewater, which were more prominent in the European north).[29] In England and Continental Europe, ambergris transformed the leather industry; its strong scent was perfect for leather dressings, as was its greasy form. Though the link between ambergris and leather was integral to the widespread appeal of scented gloves, most scholars cite the influence of Italian courtiers on either Catherine de Medici or Elizabeth I as inspiring the trend.[30]

In discussions of continental fashion, scented gloves were one of Catherine de Medici's many revolutions in style.[31] She stunned the sixteenth-century French court with her heavily scented gloves, which were then a popular fashion among Florentine courtiers. It is said that she was so fond of her scented gloves that she brought her personal perfumer, Renato Bianco, to France with her when she married Henry of Orleans, later King Henry II of France. Bianco opened a thriving shop in Paris and sold perfumes, cosmetics, and poisons near Notre Dame in the mid-sixteenth century.[32] The trend for scented gloves in Paris continued to grow in the second half of the seventeenth century. By 1656, the Parisian guild of glove-makers officially became the guild of glove-makers and perfumers; by the 1730s, however, it was struggling financially as new kinds of luxury goods like cotton gloves and scented fans surpassed scented leather gloves in popularity.[33]

In English sources, it is widely recounted that Queen Elizabeth I inspired the craze. The first perfumed gloves made their way to England with Sir Edward of Nevarre, Earl of Oxford, who in 1566 presented them to the queen as a token of his affection. Sir Edward had recently returned from Italy with sundry luxury items, including a pair of scented gloves. Just such a pair, gifted to the queen in 1566, is on display in the Ashmolean Museum, Oxford; approximately 16.5 inches, the gloves are white kid, with gauntlets embroidered in gold. They most likely were perfumed. In most accounts, the Queen, known to admire her long, slender hands, was quite taken with them and wore the Earl's gloves incessantly, inspiring the trend for gloves scented with "the Earl of Oxford's perfume."[34] By the 1590s, the trend for "Oxford" gloves was well established. In a letter dated December 30, 1592, Dr. Thomas Holland of Exeter College, Oxford, wrote that he sent William Cecil, Lord Burghley, a pair of Oxford gloves for a New Year's gift.[35] Elizabeth I was not the only English ruler to appreciate these luxury items; James I received gloves from the Spanish ambassador in London, and the Prince of Wales, later Charles I, on a trip to Madrid in 1623, spent 16,846 reales, or 21 percent of his clothing expenditures, on gloves.[36] Likewise, Charles I and even Oliver Cromwell possessed expensive scented gloves.[37]

Elizabeth I's gloves on display at the Ashmolean were most likely given to her as a New Year's gift.[38] In a royal procession from 1577 to 1578, she was presented with five pair of perfumed gloves; in 1588–89, she was presented with three.[39] She also routinely gave perfumed gloves as a reward for loyal service.[40] During a 1575 visit to Dr. John Dee, in Surrey, she willed him "Keeper of the Gloves to her Court," allowing him to keep her pair of gloves, and those of her privy chamber, so that he might "give [them] to her to wear when she was there."[41] Similarly, Hilliard's miniature of George Clifford, third Earl of Cumberland, depicts him wearing a glove fastened prominently to his hat along with an ostrich feather, presumably referencing a pair of gloves given to him by the queen for service against the Spanish Armada.[42] Ironically, the glove that literally served as the feather in Clifford's cap—a symbol of England's defense against Spanish invasion—was most likely perfumed in the Spanish manner.

The queen's Spanish scented gloves reveal the importance of luxury goods and the complicated political and economic networks that fueled their exchange, including a wide variety of sixteenth-century licit and illicit economies.[43] In 1598, Robert Cecil, secretary of state and the Earl of Salisbury, wrote two letters to Thomas Edmondes, the queen's ambassador to France and the Netherlands,

Queen Elizabeth I's Gloves (1566). Ashmolean Museum, University of Oxford

begging him to send perfumed gloves for the queen. In the first, he writes, "I have thought good to further ask you by this letter to do me so much courtesy iff possibly you can to secure me som Spanish gloves of the same perfume of which you sent to Bishop Stanhop." In the second, he writes, "[The Queen] is much pleased that an Englishman (as she sayeth) had the witt to gett any good thing from a French man. These which you have sent, are of two sorts, and so I desire black and white gloves, as I have often seen out of Spaine and Portugall, but they be oyly and ill favoured. If you think that in France any such be to be had, I pray you in any wase send me some whatsoever they cost you and I will answer it here where you shall appoint me."[44]

Though Cecil decries the oiliness of the greasy Spanish gloves, the queen values them, praising both Cecil and Edmondes for their wit in procuring them from the French. Given that Cecil's request was articulated in the midst of his inquiries about the threat of Spanish invasion through Ireland, the queen's desire for luxury items almost acquires a political urgency. Gloves most likely

Nicholas Hilliard, *George Clifford, 3rd Earl of Cumberland* (circa 1590). Though there are many feathers in Hilliard's cap, the prominent glove indicates the Queen's recognition of his service in defeating the Spanish Armada. The irony is that such a glove was most likely scented in the "Spanish" style. F6479-001, Photo © National Maritime Museum, Greenwich, London needed to be procured from France since trade with Spain had become increasingly difficult.[45] That an English queen asked her ambassador in France to procure Spanish gloves indicates their value in the late years of her reign.

Producing Scented Gloves: London's Guilds

Perhaps inspired by such royal fashion, early modern guilds produced huge quantities of scented gloves in London during the second half of the sixteenth century. Though the leather sellers guild officially regulated the importation and sale of perfumed gloves (since glove making was one of the lesser skills of the tanning trade), glove making, as an economic enterprise, required different skilled artisans, including skinners, tanners, tawers, cutters, dressers, embroiderers, grocers, apothecarists, and merchants. Scented gloves could be procured from a variety of city guilds, including the leather sellers, haberdashers, shoemakers, milliners (or purveyors of Milanese and Italian goods), grocers, apothecarists, and glovers. The complicated chain of production, combined with the multiplicity of trade guilds selling perfumed gloves, made it difficult to regulate abuses in the trade regarding both the quality of the leather and the scented dressings. Both mattered greatly to consumers but were difficult to verify.[46]

Tanning leather took time, upward of at least a year; by the sixteenth century, most skins were tanned outside London, near riverbanks, due to the noxious smell associated with this lengthy process.[47] Gloves were thus always scented objects within London, though city officials sought to minimize this fact by coding such production as "noisome" and removing it from the heart of the city.[48] Skins were first soaked in water, then covered in lime paste and slaked in pits for weeks at a time to remove all hair. Once that process was completed, the skins were scraped and washed again so that any defective leather could be removed. The skins were washed a third time and soaked in a solution of warm water, ammonia, and manure in order to render them soft. They were washed again and then placed in a fermented solution of wheaten flour of bran, which caused them to swell and become pliable. Finally, the leather was rinsed and scraped before being were dressed in fish oil (usually cod oil), alum, or vegetable extracts.[49] Only then was leather pronounced "in the white" and ready for sale.[50]

Several shortcuts could hasten this process, some producing finished leather in less than three weeks.[51] Such leather appeared tanned, but it would quickly crack under normal duress. If sold by candlelight at night, consumers could not verify that leather products were of high quality.[52] Though there were numerous bans on selling leather at night, it was impossible to monitor the practice. Given the abuses, perfumes offered one way for consumers to (at least temporarily and in the absence of sight) identify quality gloves. Perfumes also solved a practical problem involved with tanned leather. Undressed leather in its raw form emitted a distinct and horrific smell.[53] And, given the use of lime, manure, alum, fish oil, and vegetable resin in the tanning and tawing processes, dressed leather smelled strongly as well. Ambergris, when combined with other animal-based perfumes like civet and musk, was strong enough to counter the production smells, thereby revolutionizing sixteenth-century leather industries. A glove's dressing thus became an important indicator of value, particularly if it was dressed in the Spanish style.[54]

Though the queen's gloves may have been made in Spain, most gloves dressed in the "Spanish style" were not imported, nor were they made with imported leather. Rather, they were made with English skins. All imported gloves were banned in England, as was Spanish shoe leather.[55] London leather sellers, in fact, advertised all gloves sold as "London town-made gloves."[56] Though perfumes in the Spanish style were highly desirable in English markets, actual Spanish gloves were not. The Spanish Company, a joint-stock company formed in 1577 to foster trade with Spain and temporarily disbanded in 1584 due to war, never emerged as a possible commercial source for luxury gloves.[57] The smell of London's leather industries, particularly scented leather, thus offers a way to understand how an emerging discourse of nationalism grafted itself onto the body and the body politic. Yet it also reveals the extent to which certain smells permeated London and how others came to define distinct neighborhoods, docks, borders, and suburbs. The guild wars that emerged around the production and merchandizing of ambergreased gloves document the important role of smell in defining London's market spaces and, ironically, in defining (if in the negative) the meaning of "English" skin.

English animal skins were to be used for "Spanish style" gloves. Leather was one of England's largest exports, second only to wool, and London was the center of this industry.[58] London leatherworkers were concentrated south of the Thames, in St. Giles, Cripplegate, and the liberties of the Clink (including Southwark) near spice markets and Portingale communities, who potentially could trade in ambergris. Prior to the sixteenth century, leather industries were concentrated in England's rural areas, environments well stocked with deer and sheep like Perth, Dundee, Yeovil, Woodstock, Chester, Worcester, and Limerick.[59] As the leather industry became specialized, only the largest population centers could support the artisans needed to produce such goods, including skinners, barkers, "whittawyers," cutters, dressers, and merchants. As London's population grew, so too did its supply of skins, mostly from animals consumed for meat.[60] London also had the artisans necessary to support a growing demand for luxury leather goods, including those who worked in the lesser trades, like glovers and shoemakers.

Because artisans were few in number in the early sixteenth century, only one guild was required to regulate the trade in leather goods: the Worshipful Company of Leathersellers. Formed in 1533, the company was relatively new, emerging as part of Henry VIII's economic restructuring of London's guilds.[61] By the 1560s, however, the guild had become plagued by accusations of rampant abuses in the trade: cutters routinely complained that sellers mixed quality leather with inferior skins. Parliament intervened in 1563, banning the exportation of English skins and regulating tanning, curing, and shoemaking through random searches.[62] Though the statue purported to protect the reputation of London's leather abroad, it targeted domestic leather products sold in London proper. In an attempt to dispel such regulation, the Leathersellers' Company increasingly identified inadequate skins as foreign, linking abuse in the trade to anxiety about foreign merchants (and merchandise) within London's gates.

Luxury goods, like gloves scented in the Spanish style, complicated such distinctions; though produced with domestic skins, such gloves accrued luxury status through the application of foreign perfumes, often imported by foreigners in London. By the late sixteenth century, luxury goods like Spanish-scented gloves made up a large component of London's markets, so much so that many deemed them responsible for creating a dangerous reliance on foreign goods, as well as on foreigners in England's midst.[63] As early as 1549, Sir Thomas Smith, a London haberdasher and skinner, argued against the proliferation of luxury goods, contending that England was "overburdened with unnecessary forrayn wares" and should either ban them or "els make them within oure owne realme."[64] And William Strafford, writing on the decay of trade in Coventry, noted in 1581 that "there is no man can be contented now with any other gloves than is made in France or in Spain."[65] By the 1570s, there were approximately ten thousand foreigners in London, making up about 10 percent of the population. Companies like the Leathersellers' used statutes designed to regulate domestic trade, such as the requirement to perform random searches, to target foreign workers in London.

Other aspects of the trade, specifically glove making and leather dressing, remained unregulated until 1592 when the discourse of abuse began to infiltrate the lesser trades. In 1592, Sir Edward Darcy petitioned Parliament to extend the regulations instituted in the 1560s on the heavier aspects of the industry to the lighter leather crafts. Again citing widespread abuse of the trade, Darcy's petition accused the Leathersellers of mixing inferior "deceiptfull taweing of base and unserviceable leather" and sought to institute random searches of Leathersellers' wares in an attempt to protect glovers from the inferior leather materials sold to them by other members of their guild.[66] Darcy's petition failed since it directly interfered with the Leathersellers' charter, which assigned the right to search only to guild members, but it demonstrated an important shift toward public regulation that would resurface in James I's reign. After a frustrated attempt to form a separate guild in 1619, the cutters and dressers would finally reincorporate themselves as the glovers' guild in 1638.[67]

The leather industries sought to limit those outside traditional guild systems or those who ignored regulations (by taking on too many apprentices, including women).[68] Yet those involved in London's nascent perfume industries, including grocers and grocer-apothecarists, depended on foreign or illicit trade networks.[69] Between 1600 and 1620, the values of imports had increased by 40 percent.[70] Because the value of perfumes rested on their imported status, those who sold aromatic ingredients could not easily argue that their products were domestic. Rather, the struggle over perfumes turned on the difference between medicinal uses of spice and commercial, culinary, and sartorial ones. Attempts to regulate glove production in London demonstrate the extent to which olfaction challenged traditional approaches to urban regulation. Maps like Roberts's offer a compelling glimpse of London's economic realms, but their smellscapes were more difficult to control than he lets on.

As my explorations of the medicinal use of sassafras in new world contact zones in chapter 3 and the use of pomanders in plague-infested city realms in chapter 4 demonstrated, scents were powerful medicinal curatives that could saturate a space with their smell. But, as the fashion for scented gloves grew more popular, perfumes were, once again, associated with luxury. Ambergris was almost exclusively a luxury perfume ingredient, with almost no known medicinal applications. It was, however, imported along with other spices, like cinnamon, ginger, pepper, and galingale—spices with medicinal and culinary uses—by grocer-apothecarists. In the early sixteenth century, the College of Physicians, incorporated in 1518, was in charge of training and regulating the practice of medicine in London. Doctors were trained in diagnosing diseases and prescribing remedies, which were then filled by grocer-merchants or grocer-apothecarists. The grocers' guild, one of the oldest and richest in London's history, controlled the importation of food and spice, including valuable aromatic ingredients used in medicinal prescriptions. The apothecarists were thus in a precarious position between two powerful organizations, the College of Physicians and the grocers, and the medicinal and commercial uses of aromatics.

The emergence of the apothecarists as a separate guild thus required complex negotiations with both rival companies, which came to a head at the end of the sixteenth century. Disputes with both guilds were not new: the apothecarists' tenuous relationship with the grocers had existed almost as long as their trade.[71] Almost immediately after the grocers' charter was granted in 1303, for example, apothecarists began to complain about others encroaching upon their business, particularly London's foreigners. In the early fourteenth century, the Carta Mercatoria's recognition of strangers' rights to sell lucrative electuaries and compounds of the apothecarists' trade made them wary of the guild's role in protecting their rights. By the sixteenth century, however, such complaints resonated within a growing discourse of public health, particularly focused on urban space. Rosemary, as a native herb with deep culinary roots in England, did not provoke such debate about medicine and cure; new aromatic scent ingredients, like ambergris, did. The grocers' jurisdiction over spices eroded in 1540, when Henry VIII awarded the College of Physicians the right to search the home of any apothecary and monitor the quality of goods. The physicians could not, however, destroy the faulty goods without prior approval from the grocers' guild.[72] After the 1560s, the grocers sought to protect their remaining jurisdiction over the spice trade by performing their own searches with "expert apothecaries" in their attempt to eradicate "evil wares and unwholesome spices."[73]

By the 1580s, the apothecarists' struggles were no longer with the College of Physicians but with the grocers' guild, and they aligned themselves with the practice of medicine in order to gain leverage against the grocers. Citing numerous abuses of their trade, the grocer-apothecarists petitioned Queen Elizabeth I for a separate charter in 1588, but they were denied. After virulent outbreaks of the plague in 1603 and 1606, however, such an argument gained favor, and James I incorporated the apothecarists as a separate segment of the grocers' guild. Such a strategy was part of James I's broader initiative to reorganize and reincorporate London's guilds under the aegis of public health. Warnings about abuse in the trade, particularly of the sale of "faulty" and "evil" medicines, gained strength.[74] The debates about reincorporation turned on the right to search—and destroy in fiery public displays—rivals' wares. By 1617, the apothecarists were granted a separate charter by the king. In order to understand how an ephemeral scent like ambergris could have such a profound impact on London's economies, it is necessary to delve into the material histories of its guilds.

Guild charter regulations were anything but precise, due to the plethora of new ingredients (like ambergris) with confusing, or multiple, origins. Consider the tautological language from the apothecarists' charter defining the mystery of their art and trade: "Medicines, simple or compound, Wares, Drugs, Receipts, Distilled Waters, Chemical Oils, Syrups, Conserves, Electuaries, Pills, Powders, Troches, Oils, Ointments, Emplasters, or any other things whatsoever which belong or appertain to the Art or Mystery of Apothecarists." The charter, however, granted the apothecarists the right to search and regulate abuse in their trade. If any of these "things whatsoever" should prove "unlawful, deceitful, inveterate, out of use, unwholesome, corrupt, unmedicinable, pernicious, or hurtful," the apothecarists' charter preserved the right to immediately burn said item "before the Offender's Door," ironically creating a dangerous scent and releasing it directly into public spaces.[75]

Moreover, charters document that the destruction of faulty medicines were to be done in a "public" manner, which raises fascinating questions about what it meant to consume aromatics in the period and to navigate London's market spaces.[76] Searches occurred three times a year and both the grocers' guild and the apothecarists' records document such seizures, recording the destructions of many barrels of "faulty" treacles, stramony, and sandalwood powders in fiery and public demonstrations.[77] For example, in July 1610, grocers Nick Warren and John Allen were "complained for selling unwholesome Treacle, and were committed to Newgate"; in December 1612, from the shop of "Lewes Lamtre, apothecary in Lymestreet London the quantity of iij [3]oz of stramony" was declared as "defective wares" and destroyed by fire.[78] Sample records continue, listing only the date and the amount of goods destroyed: "On July 22, 1614, 8 barrells of Treacle, to be destroyed"; on December 16, 1618, "9 Barrells" of "defective triacle heterofore taken in the search and yet remayning in the house were adjoined to be consumed and burned in the garden which was done accordingly."[79] These searches continued well after the establishment of the apothecarists' charter.

Thus, under the auspices of regulating abuse of the trade in scented goods, members of all three organizations entered one another's shops, attempting to purchase spices, simples, oils or other aromatics from inappropriate purveyors (i.e., members of another guild). Once sold, the goods were declared "faulty" and were immediately seized and burnt before the merchant's door, or they were collected and transported to the guild halls for a massive burning of faulty medicines.[80] Though there were instances in which this regulation of faulty medicines actually monitored and caught defective medicines, the frequency of seizures suggests that the discourse of abuse in the trade was central to controlling the lucrative sale of aromatics. Like in the forests of the new world, fire proved an effective way to extend a scent into space, redefining and changing its material environment.

Both guilds appealed to the monarchy for relief. James I assigned specific spices to the appropriate guild: those relevant to medicine were the domain of the apothecarists; all others were the domain of the grocers. Yet the line between aromatics and medicines was almost imperceptible. In 1623, five years after the incorporation of the apothecarists' guild, an order by the king required each guild to provide a schedule of ingredients that made up the bulk of their trade. The king once again sought to create disciplinary divides through the assignation of spices. Not surprisingly, the guild schedules are almost identical, with the exception of six ingredients: arsenic, mercury sublimate, precipitator, "oyle of bay," nunmoca, and mastic. These ingredients were "peculiar to the trade of apothecary" when sold in "parcels and quantities of simples as shall be merely fit for physic."[81] The apothecarists, however, submitted their schedule in Latin; the grocers submitted their schedule in vernacular English.[82] Although this appeared to be a minor detail, the apothecarists' careful and consistent links to the medical profession in their appeals allowed them to defend their control of aromatics through an authoritative discourse of science, a discourse that established itself only after igniting noisome and hazardous smells into London's streets.[83]

The ingredients common to both guild schedules were, not surprisingly, aromatics. Ambergris, benjamin, civet, labdanum, mastic, musk, spikenard, and other general "perfumes" composed the bulk of the contested ingredients.[84] Despite the king's intervention, spontaneous searches of competitors' shops continued well into the 1630s. The rapid increase in volume and varieties of imported aromatics made the assignation of guild niches through spice ingredients nearly impossible. These tales of covert purchases of spiced goods, public seizures of faulty medicines, and public bonfires of exotic ingredients remind us that medicine was a fiercely contested economic and disciplinary domain in late-sixteenth- and early-seventeenth-century London and that olfaction was a key component of an emerging discourse of public health. It is easy to see why scented gloves with ambergris, as luxury items, became a coveted item of desire in a market where almost every other scent had a use or purpose that was not solely pleasant.

Although the apothecarists' guild struggled for solvency throughout the first half of the seventeenth century, by the early 1630s, its members were able to purchase their first guild hall; by the 1670s, they were able to rebuild their Great Hall, form the city's first dispensary, and open the city's first garden of physic in Chelsea. By the late 1670s, their presence and stature as dispensers of medicines and medical knowledge began to challenge the physicians' prominence. The history of the apothecarists' struggles with the Royal College of Physicians is well documented; their remarkable ascendancy as a solvent and formidable livery company, however, matches a near reversal of power of the grocers' stature within London. This chiasmus-like transference of power registers the economic importance of perfume ingredients and the luxury goods doused in them in the period.

The apothecarists' Latin schedule of aromatics and exotica was an important departure from traditional strategies for organizing and classifying knowledge about herbs. Prior to this moment, information about both cookery and medicine was published in vernacular herbals.[85] By the 1650s, Latin increasingly demarcated health professionals like surgeons, physicians, and apothecarists from other health practitioners, like housewives, herb-wives, and midwives. Yet it also revealed that perfumes had other commercial applications. Guild histories provide one perspective on how the fad for scented gloves resulted in control and regulation of leather and spice industries. But the references to perfumers in London's alien registers and to perfumes in women's recipe books complicate this history, revealing that many more Londoners were "all oiled in ambergreece" than those appropriately sanctioned in guild warden's records.

Selling Perfumes: London's Luxury Markets

In the final chapter of *Theatrum Botanicum* (1629), a mammoth study of the earth's plants, John Parkinson, herbalist and apothecary to the king, turns his attention to exotica and "strange and out-landish" plants.[86] Parkinson's lexicon marks the apothecary as an appropriate point of entry through which aromatics should flow, transforming imported ingredients into medicinal drugs. Describing himself like a "Fallen Adam" from a "paradise of pleasant flowers . . . into a world of profitable Herbes and Plants," Parkinson outlines how "outlandish" plants become so called "spices and drugges in our Apothecaries shoppes."[87] Most of the "outlandish" plants in Parkinson's final section are aromatics. Sorting odiferous herbs, plants, and spices by their relationship to English soil, and, by implication, their market uses, Parkinson, participates in broader cultural concerns over the origins of imported aromatics and their effects on English bodies. Parkinson's focus on the nation as potentially penetrated by smell echoed early modern gynecological conceptions of breathable wombs also penetrated by smell.[88] His treatise ignores altogether the possibility that such ingredients could become luxury perfumes, or that such penetration might be pleasurable. By 1629, however, readymade commercial perfumes were widely available in London; such ingredients had another recognizable point of entry: the perfumer's.

The development of the occupation of perfumer underscores how olfaction came to be linked with pleasure. Given the ways certain industries associated with noxious smells, like leather-tanning, were pushed out of the city's centers, the labor of perfuming (and its smells) offered contradictory meanings within it. As the guild history charted above demonstrates, one could purchase perfume ingredients from a variety of sellers. One could not, however, purchase perfume from a perfumer until after the first quarter of the seventeenth century; before then, one hired a perfumer to air a room. Perfume was thus a verb and not a noun; as their name suggested, perfumers scented objects *per fume*, or through smoke. The earliest perfumers in London worked for royal apothecarists, distilling perfumes and scenting and fumigating royal chambers.[89] In the summer of 1564, for example, Elizabeth I's mistresses of the bedchamber used two pounds of orris powder (made from the root of iris flowers) to scent the room with a perfuming pan. Perfumes were also used to air the great chambers at Hampton Court, Richmond, Sheen, and Westminster, along with the queen's litter.

By the 1630s, however, perfumers were a recognizable occupational identity and a prominent presence in London's luxury markets. As perfume became a recognizable commodity, demand increased for readymade perfumes and imported aromatic ingredients. Between 1600 and 1620, the values of imports had increased by 40 percent; perfumes were potentially very lucrative, especially if one could associate them with luxury imports.[90] Early Stuart policy sought to develop luxury manufacturing at home, thereby reducing the number of imports; perfumes, as a readymade luxury commodity, were one way to do this.[91] Because the apothecarists successfully captured a part of the market for aromatics in the seventeenth century by emphasizing their use as drugs rather than spices, which were under the purview of the grocers' guild, others wishing to profit from these same imports needed a new way to market aromatics. By the time the title of "perfumer" emerged in London as a recognizable occupational identity, it tended to describe precisely those barred from inclusion within early modern guilds— early modern alien men and women and native women. The perfumers who thus sold scented gloves and other perfumed items in London's luxury markets offered no clues to their own provenance, cloaked in those of their wares.

Although the number of alien perfumers did not dramatically change between 1571 and 1638, their relationship to London's economic markets did as the labor of perfuming shifted from distillation to merchandizing. There were a few resident alien perfumers in London in the late sixteenth century; often, these Italian and French perfumers worked at court. For example, when Hotspur decries the effeminacy of the king's messenger in Shakespeare's *Henry IV, Part One*, he sneers that he is "perfumed like a milliner" (1.3.35).[92] (*Milliner* meant a seller of luxury goods and women's wares made in Milan.) Hotspur objects to the presence of the unnamed lord on the battlefield, suggesting that his perfume smells luxurious, courtly, and foreign, more appropriate for an Italian perfumer than a soldier. Likewise, in *Much Ado About Nothing*, Boraccio "entertains" himself as a perfumer, smoking a musky room in preparation for Don Pedro's arrival in Messina (1.3:46).[93]

Whereas the most popular scents were those in the Spanish style, or heavily ambergreased, London's earliest perfumers were Italian. London's Alien Return of 1571 lists three perfumers, indicating that the perfuming of scented gloves was secondary to the labor of perfuming performed at court. John Gillambisky, an Italian-born resident of Queenhithe Ward, is described as a "perfewmar for the most part at Court." Francis Lucatella, listed in the return as a Venetian perfumer, denizen of London, and a resident of Blackfriars, most likely worked in a similar capacity. The return notes, however, that he also "selleth and perfumeth gloves." Frauncis Jjttowe, an Italian "perfumer of gloves," who "hath bene longe in this realme," resided with one of the queen's musicians, emphasizing alien perfumers' association with court. The 1582 return corroborates that perfuming was a stable service industry. Frauncis Lectnary, who is listed as both a stranger and perfumer residing in Blackfriars, could be the Francis Lucatella who was mentioned in the 1571 return, demonstrating long-term ties to the community (and his economic success).[94] During this period, foreign perfumers were gaining economic power. Italian perfumer Gilliam Bisco, for example, is listed as a denizen of London, a patent by Parliament that mitigated some of the restrictions on aliens merchandizing in London.[95]

By 1593, most perfumers in London were Italian or French, and many were denizens with English wives, suggesting that their status was exceptional, based on their long-standing ties to London's merchant networks. Yet the return also documents new kinds of labor, six "dressers of Spanish leather" and three "aquae vitae distillers," suggesting that other professional identities were emerging in response to the fashion for perfumes. Franncis Lucattell is mentioned again in the 1593 return, where it is noted that he had an English wife and apprentice. William Bysco, or Gilliam Bisco, the Italian perfumer of gloves included in the 1583 return, is listed in the 1593 returns as French. It also reports that he had an English wife, Mary, a stranger apprentice, and that he had become a free denizen of London, a rare status granting him freedom to trade in the city without official membership in a guild. Finally, the return lists one more perfumer, a "Fraunce," who resided with John Mallert, a bachelor, in the parish of St. Peter the Poor, in Broad Street Ward, near Bishopgate—an area known for its strong ties with France.[96] "Fraunce" was twenty-four and "no denizen" of London, though he had resided there for five years and kept an English maid "of ten years." Bisco's shift in nationality, along with the eponymous "Fraunce," highlight perfume trends in the period, especially the fashion for lavender perfumes; Bisco's and Lucatell's anglicized names also suggest the assimilation of foreign perfumers into London's economic communities.

By the second quarter of the seventeenth century, perfume was a readymade luxury commodity sold in the new luxury markets of Westminster. The returns of 1627, 1635, and 1639 list six perfumers, eleven leather dressers, three distillers of strong waters, and one seller of sweet powders. Almost all perfumers were lodgers in Westminster liberty.[97] By the 1620s, perfumer was also a recognizably English professional identity. An inventory "of the Apparaile & bedding that the Adventurers have bestowed upon each of the younger woemen now sente" to Virginia in August of 1621 included a pair of perfumed "white Lambe gloves," supplied by William Piddock, perfumer.[98] English perfumers clustered in the East End. Bartholomew Benson was a perfumer of Artillery Lane; Richard Hooke was a perfumer in St. Magnus parish near Bridefoot Street, along with Henry Shawcroft and Henry Coleman of the parish of St. Botolph, near Aldersgate.[99]

Perfuming, as an industry, expanded to include new kinds of luxury commodities, as well as new kinds of professionals to merchandize them. As this happened, London's economic map of perfumery shifted from the city's central and eastern parishes (near the Royal Exchange and established French and Italian communities along Threadneedle Street) to the wealthier, western parishes along the Strand, near the New Exchange and Westminster. In the late sixteenth century, luxury goods were concentrated in the Royal Exchange. The Royal Exchange was jointly financed by foreign and domestic interests (particularly foreign and resident-alien Italian merchants, along with the Wool Staplers and Merchant Adventurer Companies), but its leases reinforced the city's guild system. As such, the sale of perfume there bolstered the broader goals of the Exchange; consumption of foreign luxury goods was encouraged, as long as it benefited economic hierarchies that rewarded Englishmen. Such regulation was not unlike those placed on scented leather goods or perfumes: it was fine to smell in the Spanish style as long as the ambergris was smeared on English skin. Though one cannot be certain that a grocer would fill his shop with groceries, the Royal

Exchange organized itself through company association, with fifty-five haberdashers, twenty-five mercers, twelve grocers, ten leather sellers, and two milliners.[100]

The Exchange sought to deemphasize its role in increased imports and imported labor by stressing its livery associations and by publicly banning itinerant peddlers from the space, including herb-wives and "bawdy baskets," or female peddlers.[101] Peddlers, like Shakespeare's fraudulent Autolycus in *The Winter's Tale,* were strangers in the community and, thus, handy scapegoats to appease anxious London shopkeepers without affecting the global financial backing of the Exchange. Autolycus, for example, sold many of the luxury items available in the Exchange, including ribbons, glass, pomanders, brooches, perfumed gloves, masks, stomachers, pins, necklaces, and "Perfume for a lady's chamber" (4.4.225, 599–600).[102] The Royal Exchange flourished due to Continental systems of credit and labor practices that promoted luxury consumption, even as it appeared resoundingly English through its reinforcement of city liveries.

There were no perfumers operating in the Royal Exchange; however, the Exchange's strong connections with Italian merchants and its location near Italian communities suggest that perfumes were available. Though we know very little about the inventories of the shops in Gresham's Exchange, the sole piece of archival evidence, Thomas Deane's inventory for his haberdashery shop, includes the type of pomanders discussed in chapter 4.[103] The shops were very small. Most were five feet by seven and a half feet and crammed with a wide variety of luxury goods, which made it almost impossible to manufacture crafts or commodities, including perfumes, in such spaces.[104]

Yet that does not mean that scent was not an important component of the sensory experience of such realms. If, as the fiery and smelly public bonfires of the apothecarists' and grocers' guilds suggest, London's outdoor market spaces were defined by their olfactory properties, then so, too, were its new, indoor luxury markets: strong aromatics would have overwhelmed such cramped, dark spaces. A preacher in Oxford utilized such second-hand smells to emphasize the power of church, even for those who only passively listen to sermons: "If any man, sayth Chrysostome upon Iohn, do sit neare to a perfumer, or a perfumers shop, euen against his will he shall receuie some sauour from it: much more shall he who frequenteth the Church, receuie some goodness from it."[105] Linking strong scents with perfumers, and perhaps their shops, this sermon utilizes what must have been a common experience (even for churchgoers in Oxford) of smelling a perfumer and his "sauour" against one's will. N. de Larmessin's engravings from late-seventeenth-century France also raise provocative questions about how perfume altered conceptions of market spaces. His engravings of both perfumers and apothecarists—literally comprised of aromatics, smoking censers, pastilles, powders, spices, and gloves—visually remind scholars working on market spaces that the environmental impact of perfume vendors could be great.[106] It is hard to imagine either perfume purveyor as offering any scent of his own, the scent of his goods permeating his body and those around him.

The association of such scents with the professional identity of perfumers strengthened with the building of the New Exchange. Completed in 1609, the New Exchange, known as "Britain's Burse," focused almost exclusively on exotic, and imported, luxury items. Its location on the Strand, the main thoroughfare between Westminster and the City, was designed to entice luxury trades to drift

Gerard Valck, after Nicholas de Larmessin II, "Habit de Perfumeur" (1695–1720). Engraving. © Trustees of the British Museum

west, out of the City, and toward its wealthier enclaves.[107] It succeeded; by the 1630s, the New Exchange defined all that was fashionable about the emerging West End district of London.[108] Its covered arcades hosted a number of shops that specifically focused on imported luxury goods, particularly exotica like porcelain, feathers, and perfumes. The shops in the New Exchange were cheaper and larger than at the Royal Exchange; the highest rents were for those on the second level, particularly the corner shops. Yet, it also required merchants to focus on a specific trade, thereby encouraging the development of independent perfume shops leased to those foreclosed from London's guilds: foreigners and native women.[109]

Perfumers were prevalent on the second floor; renewal of leases from 1633 document at least three perfume shops there, including Queen Henrietta Maria's perfumer, Jean-Baptiste Ferreine, whose shop was one of the largest in the Bourse. Likewise, William Dungen, listed in the 1635 alien return as a perfumer, leased two shops together, to a Mary Jelley, "a sempstress" and "seller of French wares," and a Mrs. Hassett, "perfumer." Mrs. Hassett was most likely an abbreviation of Blennerhassett; George Blennerhassett was a neighbor of Dungen's and a fellow shopkeeper, with two haberdashery shops—the George and the Black Beak—on the lower level of the Exchange. Blennerhasset was a prominent merchant, selling many items to the Earl of Salisbury, his landlord. Mrs. Hasset, "perfumer," was most likely George's wife; her association with Blennerhasset and with Dungen would have made her a prominent retailer of perfumes to the landed gentry.[110] Mary Jelley sold "French wares," including perfumed gloves out of her seamstress shop, the Acorn. Thomas Southerne, a draper, did as well, and was fined accordingly for it.[111]

Ferreine's presence, like that of other perfumers and merchants in the Bourse, emphasized the New Exchange's royal connections. Some perfumers expanded that affiliation beyond spatial terms: Edmund Bolsworth's advertisement for his perfume shop, located at the King's Arm and Civit-Cat near Temple Bar, contained the royal stamp, as did most shops of the New Exchange.[112] Others expanded their merchandise to include snuff and other tobacco-related products; Dungen's shop names, the Phoenix and the Orange Tree, most likely emphasized that he sold both readymade scents and tobacco, or "Indian" perfume, a name that, by the 1630s, emphasized not only that tobacco was the most-recognizable scent of the new world but also conflated tobacco's perfume with natives.[113] The Phoenix's fiery classical allusions underscore the shift from perfuming to merchandizing scents, whereas the Orange Tree suggests the growing fashion for Continental scents, particularly floral essences, which are discussed in chapter 6. By the end of the seventeenth century, perfumers were also tobacconists: William Trunket, a late-seventeenth-century perfumer located near Temple Bar, sold "all sorts of snuff, perfumes, essences, French and Hungary Waters Retail."[114] Aromatics were integral to the new luxury shopping malls of the West End, linking the scent of perfume to the experience of shopping.

Consuming Perfumes: Housewives

Perfume, as a term, connoted both an act of labor and a commodity product; thus, the work of perfuming could mean different things depending on who was producing and consuming it. For alien perfume makers, the work of perfuming often meant producing readymade, luxury perfume commodities such as distilled waters, scented powders for linens, or perfumed sundries, like gloves or ribbons. Their labor of perfuming included merchandizing the commodity of perfume. For housewives, the labor of perfuming mimicked general fashions for scented goods but, when performed in the home, produced very different scents, depending on the quality and quantity of ingredients. Even luxury scents could be created at home. Gloves scented in the Spanish style, for example, required maintenance to preserve their fragrance. Recipes from the period reflect the same ingredients most likely used in the composition of readymade commodity perfumes, but they also reveal both individual preferences and market availability. For example, one recipe calls for "damaske" roses, whose "sent" is "as good as you can find"; this advice reinforces the superiority of the damask rose scent yet admits that lesser varieties might be used.

As Parkinson advised in *Theatrum Botanicum,* aromatics could be profitable if they were marketed as "spices and drugges."[115] Once in the home, however, the artificial distinctions between cookery, medicine, and luxury eroded.[116] A wide variety of domestic labor was performed in early modern kitchens, and, increasingly, new types of ingredients were required to perform such tasks.[117] Analyzing representations of wives as domestic housekeepers, Natasha Korda argues that early modern domestic manuals sought to instruct women in the proper consumption of foreign luxury goods "infiltrating the home."[118] These manuals also document that housewives knew how to produce luxury goods at home. As women learned new ways to consume aromatics, they also learned new ways to produce luxury goods, especially perfumed gloves. Print and manuscript cookbooks document that housewives knew how to dress the leather so that they retained such scent.

Targeted toward a mostly female audience, most published seventeenth-century domestic manuals market themselves as part of women's secret knowledge. The epistle to Sir Hugh Plat's *Delights for Ladies to Adorne Their Persons, Tables, Closets, and Distillatories; with Beauties, Banquets, Perfumes, and Waters* (1609) invokes an absent perfumed manuscript: "But now my pen and paper are perfum'd, / I scorne to write with coppresse or with gall, / Barbarian canes are now become my quils, / Rosewater is the inke I write withal."[119] Domestic knowledge, made up of odiferous herbs, drugs, and spice ingredients, is imagined as perfume. *Delights for Ladies* includes over fifteen recipes for perfumed distillations, mixing England's native ingredients of gillyflowers and roses with imported spices like cinnamon, civet, and sandalwood. Five different recipes describe how to produce rosewater, along with recipes for "sweet powders, ointments and beauties."[120] There is also a recipe for perfumed gloves. One of the few in the collection to include ambergris, the recipe can "sufficiently" perfume up to eight pairs of lambskin gloves, with adjustments should one wish to perfume kidskin or goatskin gloves.[121] The instructions suggest the extent to which women consumed leather gloves and indicate that consumers of perfumed gloves know how to redress the leather at home, infusing women's kitchens with the scent of luxury.

Although Gervase Markham's *English Huswife* does not include any recipes for cosmetics, it does list recipes for perfumes, smokes, pomanders, sweet waters, and scented gloves.[122] Similarly, Margaret Yelverton's 1621 "Booke of Phisicke Surgery Preserves and Cookery with Sundrie Other Excellent Receites" contains recipes for medicines and other instructions useful for maintaining both health and domestic life, like fumigating a sickroom.[123] It also contains several recipes for producing scents if commercial ones are unavailable or too expensive. To perfume apparel, for example, she advises to keep it clean and "perfume it often either with some red powder burned or with Juniper."[124] Likewise, to make sweet water, she recommends using whatever sweet herbs are available: "take all maner of sweete hearbes and pottle or gallon of faire water, fill yt till you have out all the water, fill yt againe and fill the styll with rose and beaten cloves, put to yt and fill it till you have out all the water." Similarly, she provides a recipe "to make sweete bags with little cost."[125] Mary Dogget's book includes recipes for pomander-bracelets, a perfume for a sweet bag, another for "sweet bags and sweet water," and one to perfume gloves.

Mrs. Hughes's manuscript commonplace book, penned as the "receipts of her whole booke, written in the year 1637" includes a translation of an entire book of recipes for Spanish perfumed gloves.[126] Her manuscript translates over seventy-six recipes for producing perfumes, including fifteen specific recipes for perfuming gloves. She concludes with a nod of deference to those with more "expert" knowledge but ultimately chooses to excise those ruminations from her translation: "But for all this, the expert Apothecary in the main matter, in whome, or in whose Experiments, consists the key of this work; I would sett downe the conditions hee ought to have, but that I will not bee too tedious; or make a little sum divided into this parte." This final postscript, written just before the index, suggests that the she has an eye toward other reader's interest in and use of her text. It also reveals that the apothecary's shop was not the final destination for spices imported into England. When spaces like Mary Jelley's shop in the New Exchange or Mrs. Hughes's kitchen are included in London's economy, a new map emerges, one that connects the city's interior luxury markets and domestic kitchens with its open-air manufacturing sites and public streets through the smell of ambergris.

It is not surprising that housewives quickly learned how to reproduce the sundry powders, pomanders, sweet waters, and perfumed objects produced by grocers, apothecarists, perfumers, glovers, milliners, and haberdashers. Furthermore, their recipes reveal a synthesis of consumer products with practices designed to create a more personalized product. Early modern women's kitchens represented a microcosm of England's luxury markets, defined on a smaller scale by many of the same pleasant and noxious smells that marked those public spaces. Over three hundred recipes for perfuming gloves appear in over sixty distinct manuscript cookbooks dated between 1580 and 1640; forty of these books are identified as women's recipe collections. Many of the recipes, like the one I quoted earlier, are translations of popular printed English or Spanish cookbooks, further emphasizing the ways perfume participated in the construction of private space and public, and national, identity.[127] For example, Mary Doggett's recipe for gloves perfumed in the "Spanish" manner directs women to perfume their gloves until they "swim" with amber and emphasizes the "Spanish" ingredients of ambergris, civet, storax, and musk.[128] Another manuscript cookbook contains a recipe for "How to perfume gloves as be made in Spaine and Portguall and translated forth of a written Spanish book of the duches of Ferius," which directs one to anoint the gloves with ambergris many times till they "may drink upp a great part of the said ointment."[129] Once the gloves have "drunk up the ointment," one must "take the gloves and lett be very well rowled up in a faire paper very close that they loe not their smell and lett them lye three nights layerd under the first quilt of the bed you lye on."[130] Instructions like these demonstrate that, although the scent of ambergris might be ephemeral, its material impact on English economies left a distinct trace. Furthermore, the recipe does not register the same anxiety about foreign perfumes on English skin, collapsing gloves in the Spanish style with gloves made in Spain and Portugal.

By the end of the seventeenth century, published cookbooks explicitly targeted new audiences, specifically women who lived farther from the capital's shopping districts, providing them with the means to create their own fashionable perfumed items, without "extravagant" cost. Barbe's *French Perfumer,* for example, targets "Persons of Quality and Condition" who can "afford themselves leisure enough to gather Flowers at their Country Seats, and make use of them for Perfumes," thereby not only "diverting" them but also saving them "the Expence of buying them at Extravagant Rates in Shops."[131] Many of these manuscript cookbooks are composite collections of recipes, including ones copied from print sources and original recipes developed by anonymous authors. Most declare themselves, at least in their origin, to be the work of women. Some are fair copies; most are not. Read together, they reveal that housewives participated in producing their own luxury goods.

If perfumed gloves are any indication of other types of luxury items, then a greater variety of Londoners produced and consumed such goods than we have previously thought. The economic struggles to control luxury aromatics used in perfuming leather gloves drew sharp distinctions between medicinal, culinary, pharmacological, cosmetic, and sartorial uses of aromatic ingredients like ambergris. By the end of the 1630s, perfumers emerged as a distinct professional identity and perfumes as a recognizable commodity. Many early modern Londoners were, like Lady Davies's angel, "all oyled in ambergreese." Her inclusion of this small detail within her broader tale demonstrates that gloves were more than the sum of their visual parts; scents were crucial to interpreting their social and economic worth in the period. Lady Davies probably knew that many of her female readers would have a rudimentary understanding of scents; while the general consensus may have been that she was mad, the olfactory details of her vision depict her angel, and by association herself, as a very fashionable figure.

As these economic histories suggests, when one switches from visual to olfactory registers, new maps of London luxury markets emerge, raising interesting questions about how consumers experienced them. Tales of covert purchases of exotic spiced goods, public seizures of faulty medicines, and public bonfires of exotic ingredients demonstrate that London's outdoor market spaces were defined by aromatics. Likewise, the tinctures, pastilles, and perfumes sold in London's emerging luxury shopping malls and reproduced in early modern kitchens saturated enclosed, indoor spaces as well. The scent of ambergris thus wafted throughout London, connecting these zones into the scented space of the city, emphasizing to all the important role of olfaction in its realm. This growing awareness of geographical smellscapes—like Dekker's rosemary-scented sickroom or Lady Davies's ambergris-anointed prison chamber—increasingly led to xenophobic and misogynistic fears about the efficacy of such "artificial" perfumes on porous, English male bodies. As I argue in the next chapter, pleasure gardens, stocked with odiferous botanical specimens, emerged as a "natural" retreat from the clogged, polluted air of the city, which was often metaphorically described as clogged with thick, cloying perfumes. Thus, ambergris and its ephemeral, yet pervasive, materiality reminds us that early modern markets were shifting, sensorial terrains and that one needed more than merely an eye to navigate them successfully.

CHAPTER SIX

Bowers of Bliss

Jasmine, Potpourri Vases, Pleasure Gardens

In his preface to his poem "The Garden," published posthumously in Thomas Sprat's widely read *Workes of Abraham Cowley* (1668), Cowley declares that what he desires most is to be the master of a "small house and a large garden."[1] Such desire, at least according to the preface, was never fulfilled. Cowley's spatial yearning, "so strong, and so like to Covetousness," remained frustrated by his physical location. He spent most of his last days located, as he puts it, somewhere in between the city and the country. Born in London, educated at Cambridge, exiled to Oxford (with other Royalists) and then to Paris (with Queen Henrietta Maria) during the English Civil War, Cowley returned to England as a bachelor during the Restoration and ended his literary career in Surrey, where he retired to a small home and dedicated himself to the study of botany (including translating John Evelyn's *Kalendarium Hortense*).[2] Though this was better than life in the city, Cowley confesses in the preface that it was still not ideal: "I am gone out from Sodom, but I have not yet arrived at my little Zoar."[3] Cowley penned his paean to horticulture in a decidedly non-pastoral space: "I stick still in the Inn of a hired House and garden, among Weeds and Rubbish."[4]

Echoing many descriptions of seventeenth-century city life as materially and metaphorically diseased, Cowley emphasizes that he had left the "monster London," but had not yet achieved his own garden paradise, imagined as its antithesis, a pastoral Zoar to London's Sodom.[5] That Cowley imagines London as Sodom is not surprising; post plague, such references were common, especially as writers struggled to represent the invasive presence of crowds, particularly phenomenological experiences of the crowded city. Its stench was often cited in such accounts. For example, the mud holding its paved streets in place was so foul that in 1617, Orlando Busino, chaplain to the Venetian ambassador, punned that the city should be renamed Lorda, or filth, rather than Londra.[6] Likewise, Ben Jonson's "On the Famous Voyage" describes the Fleet ditch as an "ugly monster yclepèd Mud," which, when stirred, "[b]elched forth an air as hot as at the muster," like carts discharging a "merde-urinous load."[7] Given such pervasive, urban, "merde-urinous" mud, it is unsurprising that Cowley, and others like him, longed for an escape through gardening. Throughout the seventeenth century, gardens were increasingly imagined as a way to escape urban sensory assaults, promising a horticultural respite that could reduce all sensation to a "green thought in a green shade."[8]

Juxtaposed with the cramped, interior writing spaces and rubbish-strewn gardens of the preface, Cowley's poetic garden celebrates the expansive, yet ordinate, pleasures of nature. These are both sensual and literary, imagined to be like those of a "virtuous wife" who offers pleasures refined and sweet: "the fairest garden in her looks and in her mind the wisest books."[9] And the pleasures are multisensorial; the garden offers "a gentle, cool retreat," from the burning heat of the world, along with delicious grapes and melons to satisfy even the most refined epicurean. Harmonious birdsongs offer sonic pleasures, while lilies, along with the roses, visually amplify the garden's beauty. Such roses also issue forth a delightful scent that, when mingled with the scent of jasmine, creates a perfume strong enough to counter even the "pestilent clouds of a populous town."[10] Given the choice between the country and the city, as his narrator asks, "Who that has reason and his smell / Would not among roses and jasmine dwell?"[11]

Cowley's pastoral celebration of garden pleasures was not new; writers since Horace had celebrated the bucolic pleasures afforded in nature. Yet his garden—and its emphasis on olfactory pleasures within it—describes a new sensation in seventeenth-century England: the scent of jasmine. The floral scent of roses and jasmine surpasses other perfumes sold in London's luxury markets, including the ambergris-scented gloves and musk-based perfumes sold in the Royal and New Exchange. For "the earth itself breathes better perfumes here / Than all the female men or women there, Not without cause, about them bear."[12] The delights of jasmine are imagined as natural, issued forth from the earth itself, the very opposite of urban perfume worn by women (and emasculated men). Such an image reveals how much perfume had changed since the century before. Like rosemary, whose sweet scent betrayed the presence of dangerous, airborne contagions, animal-based perfumes (musk, civet, and especially ambergris) paradoxically betrayed bodily odor. Those who wear perfume "born" on their body, Cowley's narrator reminds the reader, do so "not without cause"—their sweet scents undoubtedly mask other, dangerous bodily conditions. Jasmine, imagined as a sweet breath exhaled from nature, is the very opposite of such perfumes in both its floral material composition and its dispensation. Whereas ambergris is urban and artificial, floral jasmine is bucolic and natural.

Such odoriferous pleasure was not quite as natural as Cowley purports it to be, nor as chaste; as I argue in this chapter, the scent of jasmine was a key trope for defining eroticized space in the period and one of the most prevalent ingredients used indoors for potpourri. Writing in a hired room in an inn, next to its rubbish and weeds, Cowley perhaps longed to experience the scent of jasmine exhaled from the earth itself, but, like most English men and women, probably experienced it instead exhaled from a potpourri vase. His Zoar contained an erotic, olfactory trace that hinted at the pleasures of Sodom. In this chapter, I examine how such devices and their scents redefined interior spaces as private, erotic zones through perfume. The scent of jasmine, whether it came from flowers cultivated in English pleasure gardens or from floral potpourri vases in English domestic interiors, was synonymous with pleasure, demonstrating that perfuming practices and sexual practices were very much linked. Both gardens and potpourri were designed to cultivate odoriferous pleasure. As such, perfume plays a larger role in histories of sexuality than we may have previously thought.

In seventeenth-century England, flowers were valued for a host of sensory qualities, including color taste, tactility, and, most importantly, smell; garden architecture sought to maximize such multisensorial pleasures, particularly of exotic floral specimens imported into England and valued for their fragrances like jasmine. With its small, fragrant flowers and its lush, green foliage, jasmine still remains a garden staple for edges and rows, cultivated more for its scent than visual beauty. Over a hundred kinds of jasmine species exist, almost all of which are valued for their pleasant fragrance. Jasmine, from the Persian *yāsmīn*, or "gift from god," or *Jessamine*, its sixteenth-century English form, was (and remains) a common ingredient of perfume. For example, Dioscorides described an ancient Persian perfume that was most likely composed of jasmine absolute. Likewise, its essential oil is a major component of Chanel No. 5. In the seventeenth century, three main types existed: yellow, white, and red jasmine. All three emit a powerfully unique and strong fragrance. Like damask roses in the sixteenth century, jasmine was synonymous with perfume in the seventeenth century. Unlike damask roses, however, jasmine's fragrance could only be captured through the process of *enfleurage,* a labor-intensive process in which floral petals were smeared with suet or lard to extract their essential oils without bruising the petals. This action was repeated until the fat was saturated with the scent of the flowers, making pomade that could be used as a scent ingredient or dissolved in alcohol (and diffused through potpourri vases). The appeal of jasmine-based perfumes galvanized trade in exotic floral specimens; over two thousand pounds of petals were needed to produce three pounds of jasmine absolute.[13] The resulting perfume was thus defined by both its powerful scent and its labor-intensive tactility, both of which referenced the enclosed, cultivated space of a pleasure garden.

By the end of the seventeenth century, the term *jasmine* increasingly captured a wide variety of odoriferous and gustatory pleasures, drawn from different exotic flowers from India, North Africa, North America, and Jamaica, including Asian night Jasmine, gardenias, and Japanese sumac (from which the tea is derived). It also described a host of luxury commodities including perfume, pomades, potpourris, confectionary treats, and aromatic tea.[14] The mobility of such a term linked natural, divine pleasures previously found in gardens to those cultivated, marketed, and consumed in private, indoor spaces. It also amalgamated distinct exotic specimens from both the East and West Indies into a single scent. English explorers, sailors, merchants, and botanists seeking to collect valuable floral specimens had to reckon with radically different sensory realms, but for most English men and women, botanical specimens like jasmine, imported and planted in English gardens, functioned as a floral synecdoche for places of provenance. The scent of jasmine, like the scent of sassafras, meant something very different within England than it did in the East or West Indies, yet for those experiencing it in English gardens, it was connected to odoriferous pleasure imagined to occur in such spaces.

In this chapter, I analyze the emergence of botanical scents as an important component of odoriferous pleasure. By the early eighteenth century, scents were increasingly categorized into foul or fragrant smells. Floral scents in Linnaeus's *Philosophia Botanica,* for example, were either "ambrosial" or "noisome." While foul-smelling odors (like those associated with disease, as analyzed in chapter 4) were regulated under the aegis of public health, fragrant scents and perfumes increasingly became associated with private realms. Historians of hygiene mark a profound shift in the seventeenth century away from heavy, animal-based musk toward fragrant, "natural" scents of botanicals.[15] As such, the perfumes from English pleasure gardens and potpourri vases exhibited a delicate balance between nature and artifice. Pleasure gardens and potpourri vases were designed to create multisensorial zones of pleasure. Jasmine, in particular, represented a cultivated luxury that resulted from scientific advancements in horticulture and circum-Atlantic trade networks. An exotic flower grown in both the East and West Indies, jasmine was difficult to produce in England; its scent made it a highly prized specimen in private greenhouses and gardens. Potpourri sought to recreate jasmine's cultivated luxury (and the pleasures it inspired) indoors as well as out. Smoke perfumes, potpourri, and other means of scenting airs document how perfumes helped redefine the bedroom as an enclosed, erotic zone like that of the pleasure garden.

In many ways, the history of the English pleasure garden is a literary history. During the seventy years between the writing of two of the most sensual literary gardens in English Renaissance literature—Spenser's bower of bliss and Milton's Eden—England experienced a revolution in gardening practices.[16] As early as 1560, a new type of garden—one cultivated solely for exhibiting exotic specimens and experiencing delight through sensorial engagements with nature—became popular: the pleasure garden.[17] Seventeenth-century pleasure gardens, like early Tudor damask rose gardens analyzed in chapter 2, were designed to cultivate odiferous pleasure. Whereas the scent of roses (even foreign specimens like the damask rose) in Henry VIII's royal pleasure gardens reconfigured monarchial power through erotic proximity to the kingly body, emphasizing an English national identity, the scent of jasmine offered something very different. By the end of the seventeenth century, the scent of jasmine was powerfully linked to exotic pleasure, defined on jasmine's foreignness. Such exotic and erotic encounters with nature were increasingly staged through the scent of jasmine, linking pleasure and reason to planned zones of pleasure. Put another way, there were practical and material ramifications to the rhetorical question Cowley posed in "The Garden": "Who, that has Reason, and his Smell, Would not among Roses and Jasmin dwell?"[18] Dwelling among jasmine became possible through the technologies of gardens and potpourri.

Bowers of Bliss: English Pleasure Gardens

The fact that such gardening practices evoked the scent of pleasure rather than the sweat of labor remains a historical irony integral to understanding how literary metaphors came to define urban and rural space. As Raymond Williams famously argued in *The Country and the City*, the literary gardens celebrated and depicted in country house poems of the early seventeenth century depended on the erasure of real-world gardening practices.[19] While these evocative poetic descriptions remain a testament to the importance of gardens and gardening practices to the architecture of seventeenth-century English homes, unearthing the material history of such practices has proven difficult. The field of garden history has emerged, in the words of one critic, as "a relatively neglected window into the past, capable of recapturing much that would otherwise be lost or remain nebulous in the areas of literature, history of ideas, social and economic history, horticulture, biography, politics—the list goes on."[20] That imaginative window promises a glimpse at understanding seventeenth-century public life.

As cultivated, enclosed spaces, gardens are also central to understanding the emergence of private life. Modern historical approaches usually seek to penetrate this sphere visually—to "open" this window into the past, re-creating these spaces based on broad cultural theories of landscaping in the period or a specific gardener's description of a particular garden in the past.[21] Due to such evidentiary pressures, most scholars of garden history cite the early eighteenth century as a critical moment in English landscaping traditions in which a distinctly "English" style emerged.[22] This style revised earlier continental garden layouts, emphasizing horticultural experiments that cultivated "innocent pleasure" in nature's designs in ways that were pleasing to the eye.[23] Though such spaces were defined as "English," much of the plant matter that filled them was not. Many of these horticultural experiments grafted exotic specimens to native English flora, alarming some who worried that such experiments "denatured" nature. The poetic narrator in Andrew Marvell's "Mower against the Garden," for example, complains that the "enforced" layouts of these gardens distract from "willing nature's innocence," imagined to be dispensed as a "natural fragrance." The irony is that odoriferous pleasure was often the most important part of a pleasure garden's design.

It was thus important that the scent of nature was "willingly" dispensed and "natural." That nature must smell "natural" may sound tautological, but it is important to remember that such spaces were defined against the dangerous and amalgamated smells both of crowded urban spaces (described in chapter 4) and of the inordinate, overwhelming (and at times intoxicating) sensory realm of the contact zones (described in chapter 3). The inclusion of foreign floral specimens like jasmine, highly valued for their scent, provided a subtle way to stage odoriferous delight without calling attention to either artifice or labor. As Amy Tigner has argued, such diffusion had broader implications for England's participation in global markets as well.[24] The flexibility of the term *jasmine* to include a range of geographically diverse plants erased the labor involved in transporting a specimen to England along with its provenance. Most people neglected to recognize that its odoriferous pleasure was linked to global systems of slavery. The term's multiplicity of meanings also focused English attention on English soil; gardening provided English men and women with a proximate sensory understanding of, in the words of one critic, the "texture" of the English countryside. Benedict Robinson, in his astute reading of John Gerard's *Herbal,* for example, emphasizes the author's profound knowledge of local space, argu-ing that it functions as an "intimate" archive of how such realms were materially transformed by the inclusion of foreign botanical matter.[25] Focusing on the En-glish tulip craze, Robinson argues that it created xenophobic fears about floral matter infiltrating England's soil. Even the name *tulip,* he notes, derives from the Turkish word for turban, highlighting the flower's eastern origins and defining it as extrinsic to England, even as it was transplanted to new climates and soils.[26]

I share Tigner and Robinson's interest in the macro- and microscopic role flowers played in creating distinct spatial zones. But whereas Tigner and Robin-son focus on the ecological and cultural impact of English consumption of (and desire for) exotic flora in English gardens, I extend this spatial analysis to in-clude the creation of interior, private space through floral potpourri, arguing that desire for intimate, sensual engagements with nature had environmental ramifications beyond garden realms. Such longing, Cowely's poetic narrator re-minds us, was multisensorial; floral perfumes, dispersed from potpourri vases, provided many English men and women a way to re-create such an intimate encounter indoors. Jasmine, with its seductive scent and wide variety of com-mercial uses, was a key part of understanding seventeenth-century spatial con-figurations of English sexuality, both outdoors and in.

The emergence of English pleasure gardens turned on a redefinition of profit and pleasure to include laboring in nature.[27] Gardening emerged as a popular activity in the seventeenth century. As more men and women experienced the pleasures of cultivating utilitarian gardens, they began to expand the horticul-tural spaces in which they worked, including new geographic realms into their designs. Merging two medieval gardening traditions—large, enclosed, pleasure parks for the pursuit of hunting game and small, enclosed gardens designed to inspire Catholic contemplation through floral allegory—early modern garden-ers made pleasure gardens their centerpieces, displacing profitable kitchen and physic gardens onto smaller plots behind or beside their dwellings. By the late sixteenth and early seventeenth century, English gardens celebrated a new kind of pleasure in nature, a pleasure that emphasized olfaction.[28]

Garden manuals provided instructions for how to cultivate pleasurable scents while cultivating profit. In addition to advice on producing "physic" and "kitchen" gardens, these manuals provided detailed advice about how to organize a garden's layout and how to cultivate multisensorial engagements with nature.[29] In *The Prof-itable Arte of Gardening* (1568), Thomas Hill locates two commodities to be gained from a garden of scented herbs and flowers: "profit, which ryseth through the in-crease of herbes and flowers: the other is pleasure very delectable through the delight of walking in the same . . . by the commodities of taking the freshe ayer and sweet smell of flowers in the same."[30] Hill's emphasis on "delight" in a garden hinges on both profit and pleasure. Whereas the former is imagined to "increase" through planting herbs and flowers, the latter is experienced through multisensorial engagement with the garden space before its floral commodities are harvested, namely enjoyment in their ephemeral perfumes, released on a sunny day.

Such descriptions reflected (and further inspired) the widespread appeal of gardening: Roy Strong notes that by the 1660s the appeal of elaborate English and Continental royal gardens had spread across the middle classes.[31] That appeal was based on the cultivation of sensory "delight" as new kinds of floral specimens from eastern and western contact zones enabled English gardeners to plant their own mini-plot of paradise.[32] These gardeners cultivated a wide variety of plants, including imported botanical specimens and enhanced native ones.[33] Furthermore, technological breakthroughs such as grafting, irrigation, and plumbing, as well as the development of the greenhouse, made it possible for early modern people to craft a unique engagement with nature. Almost as soon as there were garden manuals in print in England, they offered precise instructions on how to alter a plant specimen's shape, taste, and scent.[34]

Even in their most organized layout—the labyrinth—pleasure gardens were designed to overwhelm inhabitants with sensory input. The fact that inordinate pleasures in nature could be derived from cultivated garden zones may seem paradoxical unless one considers sixteenth- and seventeenth-century impulses to catalog and collect all of the earth's natural specimens. Private university physic gardens strove to cultivate natural specimens as pharmacological ingredients, linking macrocosmic understandings of the world with microcosmic approaches to health; gardens were developed at Zurich in 1560, Bologna in 1568, Leyden in 1577, Leipzig in 1579, Montpellier in 1598, Paris in 1597 and 1635, Heidelberg in 1600, Giessen in 1605, Strasburg in 1620, Oxford in 1621, Jena in 1629, Upsala in 1657, Chelsea in 1673, Berlin in 1679, Edinburgh in 1680, and Amsterdam in 1682.[35] These educational endeavors aspired not only to understand the earth's plants but to recover Eden itself.[36] The dream of *hortus inclusus* that fueled such collections also inspired a new kind of pleasure derived from omnipotent control over sensory engagements with nature. Whereas medieval pleasure garden layouts used flowers to create allegorical zones of Christian contemplation, seventeenth-century pleasure gardens used flowers to inspire olfactory pleasure underscoring a secular—and sensual—joy in nature.

As gardening increased in popularity, gardening manuals expanded to include, at times, hundreds of pages on precise grafting and planting strategies.

This expansion outlined new technologies, and new types of specimens, gleaned from burgeoning global trade.[37] Elaborate woodcuts of garden knots, herb borders, and labyrinths accompanied late sixteenth-century planting calendars. Seventeenth-century manuals, however, also included advice on how to incorporate grottos, fountains, and elaborate waterworks. These manuals were aimed at well-off English men and women as well as aristocratic and royal gardeners; they defined pleasure gardens to include a variety of small, enclosed, aromatic areas comprised of different floral borders, herb hedges, and fruit arbors.[38]

Gardening manuals, like other domestic treatises, increasingly presented gardening "expertise" digested for nonprofessional readers.[39] One seventeenth-century author advised, for example, that the gardening information detailed in his manual was best experienced directly: "The excellency of the Garden is better manifested by Experience, which is the best Mistress, than indicated by the imperfect Pen; which can never sufficiently convince the Reader of those transcendent pleasures, that the Owner of Complete Garden with its Magnificent Ornaments, its Stately Groves, and infinite variety of never dying Objects of Delight every day enjoys: Nor how all his Senses are satiated with the great variety of Objects it yields to everyone one of them."[40] Experience, here imagined to be a mistress, transformed a total "satiation of the senses" into garden pleasures that not only could be copied and cultivated in one's home but also offered "infinite variety" in "never dying objects of delight." Such a garden offers perpetual, and one might argue polymorphous, pleasure.

This type of infinite variety and insatiable pleasure was odoriferous: floral scents, like the jasmine used in borders and rows, defined such spaces as pleasurable.[41] The fresh air and sweet smell of flowers crafted these spaces as distinct environmental realms dedicated to heightening sensory pleasure. Although sixteenth-century gardeners certainly experienced odoriferous pleasure as a result of laboring with profitable culinary and medicinal herbs, the inclusion of new kinds of jasmine specimens in the seventeenth century transformed these encounters with nature. By the end of the century, pleasure gardens were designed to offer pleasure to those who leisurely entered their realm rather than to those laboring within them.

Early modern English medicinal experts repetitively cite the ability of fragrant air to refresh and cheer one's spirits, especially through simple enjoyment of a garden's fragrances.[42] Speculating on the "Pleasant Uses" for a garden in his *Art of Simpling* (1656), William Coles asserts the power of garden fragrances to dramatically restore the body, noting especially their ability to refresh a painter's sore eyes through engaging his other senses, particularly smell.[43] Quoting both

Francis Bacon's tale of a gentleman who resided for five days on the strong scents of garlic and onions alone and another fanciful story of a German gentlewoman, "who lived fourteen yeares without receiving any nourishment downe her throat, but only walked frequently in a spacious Garden full of Odoriferous Herbes and Flowers," Coles links garden pleasure with the satiation of appetite.[44] Unlike similar tales found in medicinal treatises, Coles's examples emphasize the geography of these scents. Merely by walking through a garden, one could inhale a wide variety of pleasant, nourishing, and restorative aromas. By the end of the 1670s, domestic fumigations were losing their medicinal applications and were increasingly linked with generating pleasure.[45]

To engage the senses, English pleasure gardens exploited the relationship between nature and artifice.[46] Jacobean and Caroline garden manuals include depictions of statues that surprised onlookers with unexpected bursts of water. Statues that urinate, spit, or lactate emphasized that gardens engaged an embodied definition of pleasure.[47] Such pleasure was both ambient and sexual. Whereas floral scents offered their own pleasurable encounter with nature, gardens' aromatic enclosures also provided a private space for sexual encounters. In both respects, garden spaces emerged as natural extensions of early modern homes, becoming "essential adjuncts" to domestic, interior space.[48]

Yet pleasure gardens were also unique spaces where conventions could be abandoned and carnivalesque pleasures celebrated. Those same fountains that lactate, urinate, or spit indicate that contrast, inversion, and surprise were important themes in early modern pleasure gardens. Such themes also extended to a garden's celebration of the senses: most gardens inverted sensory hierarchies outlined in medicinal texts. In his *Treatise of Fruit Trees* (1665), for example, Ralph Austen delineates each of the pleasures one might receive from a garden, proceeding quickly through the pleasures derived from hearing birds chirping, feeling cool leaves touch one's face, and seeing the beauty latent in the order of the planting. But he lingers on the "profits" and "pleasures" in sniffing a garden: "Chiefly, the pleasure this sense meets with, is from the sweet smelling blossomes of all the fruit-trees, which from the time of their breaking forth, till their fall, breath out a most precious and pleasant odour; perfuming the Aire throughout all the Orchard. And besides the pleasure of this perfumed Aire, it is also very profitable, and healthful to the body. Here again, Profit and Pleasure meet together and imbrace."[49] Highlighting the refreshment gained from inhaling the "precious and pleasant odour" emanating from the blossoms and perfuming the whole orchard, Austen imagines profit and pleasure to "imbrace" in England's garden spaces.[50]

Although gardens fortified all the senses, the floral perfume became a synecdoche used to describe their overwhelmingly arousing and intoxicating atmospheres. Rather than celebrate optical or innocent pleasures in nature, gardeners exploited an exquisite paradox of pleasure itself by imagining it as an overwhelming experience that results from a meticulously planned, cultivated, and artificial zone. Floral scents, for example, were deliberately juxtaposed with other types of scents, including animal-based perfume ingredients. In his emblematic exploration of the *Partheneia Sacra* (1633), or the "mysterious and delicious" garden, Henry Hawkins describes roses as a "cabinet of musks," lilies as "civet," and heliotropes as a "faint botanical anchorietess best smelled in proximity to garlic and cedar."[51] Fusing the visual allure of flowers to the "civets and perfumes [worn] about them," Hawkins concludes that botanical pleasure unfolds in a garden as if on a theatrical stage where smell is as vital as vision: "a goodlie Amphitheater of flowers, upon whose leaves, delicious beauties stand, as on a stage, to be gazed on, and to play their parts, not to see so much, as to be seen; and like Wantons to allure with their looks, or enchant with their words, the civets and perfumes they weare about them."[52] Hawkins's "goodlie Amphitheater of flowers" suggests that gardens were nature's stage for celebrating other inordinate, wanton, and enchanting perfumes.

Early modern garden manuals fostered metonymic links between the embodied effects of pleasure gardens, natural (and enhanced) botanical generation, and erotic pleasure. And more than just profit and pleasure were imagined to embrace in such spaces. One gardening manual from 1594 links a couple's erotic embrace with entwined, and possibly grafted, plants. In the first map of a garden, the couple embraces in the bottom left corner, paying no attention to the labor performed elsewhere in the garden. In the second map (printed on the adjacent page), the couple is replaced with plants. Although clearly referencing two different gardens, the knots are similar enough to create a visual link between the amorous affections of the couple and their botanical counterparts.[53] This analogy suggests that pleasure gardens served as a discrete location for erotic acts and that odoriferous pleasure helped stage it.

Bedrooms of Bliss: Potpourri

Gardens were valued for their ability to generate extreme olfactory pleasures. In the second half of the seventeenth century, potpourri vases also generated these scents indoors. John Worlidge, in his 1677 *Systema Horti-culturae*, values England's gardens for their "delightful" floral "sights and smells" and advises readers to bring them indoors: "Flowers in many things convenient are, / Our Tables, and our Cupboards we prepare / With them; and better to diffuse their scent, / We place them in our Rooms for Ornament. / By others into Garlands they are wrought; / And so for Offerings to the Altars brought / Sometimes to Princes Banquet they ascend / And to their Tables fragrant Odours lend."[54] For those who could not cultivate flowers in their own gardens, Worlidge advises cultivating window boxes, buying cut flowers from markets, or purchasing "curious Representations of Banquets of Fruits, Flower-pots, Gardens, and such like," to "satisfie the fancy of such, that either cannot obtain the felicity of enjoying them in reality, or to supply the defect the Winter annually brings."[55] Worlidge's "curious representations" of floral "banquets" matches descriptions of imported painted porcelain, especially potpourri vases that allowed even more English men and women to enjoy pleasant floral scents indoors year-round.[56] Such practices demonstrate that by the end of the seventeenth century, pleasure gardens were no longer merely extensions of domestic space but rather integral, transforming interior space.

According to historian Alain Corbin, potpourri vases first emerged from French royal fascination with the floral scents emanating from their palatial pleasure gardens.[57] Potpourri vases first arrived in England in the early seventeenth century, however, as part of wide-scale consumer desire for imported porcelain from the Far East.[58] A profitable and easily procured commodity, porcelain usefully served as ballast for ships returning from trading expeditions to East Asia.[59] English tastes for porcelain, in a sense, emerged as a historical footnote to a larger story of European desire for luxury goods, such as spices, tea, and silks, three commodities easily susceptible to water damage and thus stored above a ship's water line, leaving plenty of space for porcelain below.[60] By 1704, English importation of porcelain was profitable enough to warrant a 12.5 percent tax on auctions of all "porcelain commonly called China-ware or Japan-ware imported from the East Indies, Persia, China and other parts in the limits of the charters of the East India Companies."[61] By 1744, consumer demand had led to the opening of the Chelsea porcelain factory, England's first. By 1755, thirteen additional porcelain factories were established in England.[62]

Numerous scholars in a variety of disciplines have commented on the widespread appeal of porcelain in seventeenth-century Europe.[63] As historian Lorna Weatherhill argues, English porcelain was a relatively rare commodity used only by royals early in the century, but it was soon an indispensable tool of daily life: its appeal was widespread, cutting across gender, economic, and social strata.[64] Historian J. H. Plumb records that, in 1700, one shipment alone carried over

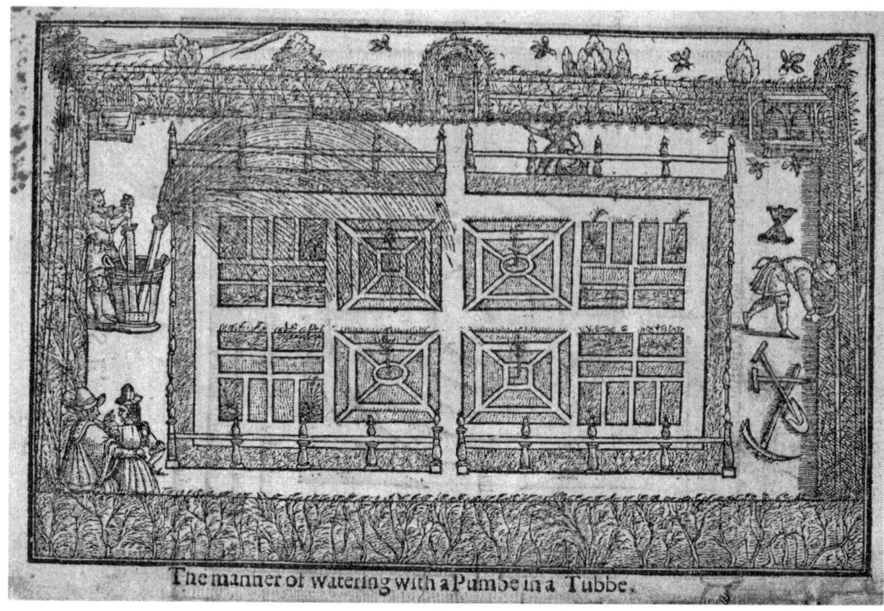

(*Above and opposite*) Anonymous, "Orchard and the Garden," (1594). Though the garden knots depicted in these two woodcuts are different, they are visually linked by the emphasis on watering "with a pumpe," whether by tube or trough. Two large potted plants suggest the embrace of the heterosexual couple. © The British Library Board, C.27.f.16, sig. G2 v, G3 r

146,748 pieces of porcelain.[65] Literary scholar David Porter describes the appeal of Asian porcelain to European consumers as a partial resistance to science's expanding domain over nature.[66] Whereas the scent of jasmine emanating from English gardens celebrated an ordered control over nature, the scent of jasmine emanating from a piece of Chinese porcelain celebrated something quite different. As Porter argues, within European colonial imagination, Chinese porcelain and European chinoiserie "emerged as a bold celebration of disorder and meaninglessness, of artifice and profusion, and exuberant surrender to all that remained unassimilated by rationalist science and classical symmetries."[67] Yet, despite evidence including vases, ship's logs, import taxes, and recipes, scholars have virtually ignored the importance of porcelain potpourri jars in seventeenth-century cultural life, and the mélange of scents they released.

The appeal of jasmine-scented pastilles, perfumes, and teas was an important part of the economic demand for imported porcelain. Early recipes for perfumed smokes combined animal-based perfumes with distilled botanical oils, espe-

cially roses, orange-flowers, and, increasingly, jasmine; by 1600, however, floral scents were used on their own, valued for their ability to craft an ambient zone of fragrance rather than to create a strong perfume meant to be worn on the body.[68] Jasmine worked especially well. As related above, the term *jasmine* captured a variety of exotic floral specimens with strong scents in the late seventeenth century, including red jasmine (or frangipane) from Jamaica, white jasmine from France, Asian night jasmine, and gardenias. Jasmine collectively represented a desire for a specific kind of olfactory pleasure in natural-smelling, ambient fragrant spaces, paradoxically achieved through structured control over nature.[69] The scent of jasmine exemplified perfectly olfactory delights cultivated in pleasure gardens. With so many potential material referents, jasmine signified the perfect perfume for seventeenth-century potpourri vases, symbolizing both a material mélange of exotic, floral ingredients necessary for potpourris and a unified, "natural" floral scent necessary for creating subtle, ambient fragrance. Jasmine perfume wafting from potpourri vases created a strong olfactory link to floral scents found in England's elaborate outdoor pleasure gardens.[70]

The lack of critical attention to the economic and historical impact of potpourri may be due to its innocuous reputation. Thought a trivial, if cloying, substance, potpourri was (and remains) associated with women's domestic spaces, imagined as one more inconsequential luxury good that defined feminine consumption in the period. Seventeenth-century potpourri vases, however, are strikingly different from what one might suppose. Most are quite large: many exceed a foot in height. Some are three feet in height. Most are round with a cylindrical base, detachable lid, and anywhere from four to eight perforations to release smoke perfumes. Although art catalogs are silent on how the vases were used, recipes in published and manuscript cookbooks (for "familiar," "cheap," or "odoriferous" smoke perfumes) contain instructions as simple as placing a few drops of essential oils on hot coals and as elaborate as burning homemade pastilles in perfuming pans.[71] Reproductions of late-seventeenth-century and early-eighteenth-century period rooms in art museums usually locate these vases in a central place in a drawing room, often next to the fireplace or on a mantel above it.[72] Those used in the dining room are usually attached to large fountains that dispensed hot liquids. Some are shaped as fish or boats, while others served only as abstract, decorative vases.[73] They are never displayed in bedroom interiors.

Their size, however, suggests the importance of perfume in understanding phenomenological engagements with domestic space, and given the association between sex and perfume in the period, it is almost impossible to imagine that they were not employed in bedroom spaces. Perfume's saturation of domestic space—both indoors and out—raises important questions about the role of gender in phenomenological encounters with ambient, eroticized space. In museums, potpourri vases are staged as part of the emergence of modern privacy and bourgeois domestic space, a space that increasingly defined public and private life through gender. Though they are often visually prominent, potpourri vases on display emit no fragrance, allowing modern assumptions about gendered, interior spaces to shape understanding of how they were inhabited. Porcelain was one more luxury commodity that defined private space through gendered consumption.

Certainly, women were the primary users of such items, but, given their size, the perfume emitted from such vases would have radically defined the space, shaping the phenomenological experience of it for all who entered its realm. This fact is clearer when potpourri—and their perfumes—are interpreted alongside the ambient, olfactory zones they meant to mimic: pleasure gardens. By the end of the seventeenth century in England, both gardening and potpourri were largely thought of as women's domains. Potpourri's commercial success was based on women's consumption of both cosmetics and luxury goods for the home. And, although botany was quickly emerging as a realm of scientific expertise, exclusive to university-trained male professionals, private household gardens were women's domain and the act of gardening, women's work. This was especially true of pleasure gardens; though outdoors, these spaces were imaginatively connected to the private realm of the home, a space that defined and expressed women's interior lives.[74] Not surprisingly, this inspired a great deal of confusion and cultural anxiety about the kinds of pleasure cultivated within such spaces. Though women gardeners are consistently represented across the fifteenth-, sixteenth- and seventeenth-century English poetry, the bucolic pleasures associated with them shift greatly, from metaphorical associations with Mary to associations with Eve.[75] Within seventeenth-century literature, gardens offer a new Eden, but women's pleasure in them, like the odoriferous pleasures emanating from them, were increasingly mapped through binaries: ambrosial or noisome.

Whether pleasure in perfume was chaste or sinful defined how one approached the neo-Edenic space of the pleasure garden. Horticultural technology enabled gardeners to produce stunning combinations of floral scents, many of which were re-created in potpourri vases indoors. Although pleasurable, these hybrid scents were often sniffed with skepticism and wariness. Literary allusions to flowers reflect a growing awareness of—and anxiety about—unnatural reproduction. Analyzing "(un)natural loving" between men, swine, pets, and flowers in Shakespeare's *Venus and Adonis,* literary critic Dympna Callaghan argues that the cultural and emblematic significance of flowers intensified in the Elizabethan period as horticultural advances blurred the distinction between "natural and artificial" pleasure. A flower's scent, in particular, marked both the allure of nature and a more sinister submission to pleasure.[76] Linked to Eden in religious discourses and travel narratives, pleasure gardens retained powerful cultural associations with sin and exoticism.[77] Jasmine-scented potpourri, however, provided a subtle solution to the problem of olfactory pleasure, allowing a mélange of exotic materials from far-flung corners of the globe to exist under a common signifier, offering a perfect potpourri for those seeking to indulge in the sweet scents of floral odoriferous pleasure indoors without succumbing to dangerous exoticism and eroticism of the garden sphere.

Olfaction suggested not only original sin but also original pleasure in nature. In Simon Barbe's 1697 perfuming manual, *The French Perfumer,* the allure of perfume originates in Eden, rather than, as his title would suggest, France: "the Origine of Perfumes is as Ancient as the Creation of the World; when the whole Earth was a delicious Garden, exhaling the sweetest Smells."[78] Linking the scent of perfume with the allure of Eden's floral scents, Barbe's ontology imagines postlapsarian perfumes through prelapsarian pleasure. This was especially true of jasmine, whose name, from the Persian word for perfume, emphasized both the flower's scent and its association with Eden, believed by many early modern English travel writers to be located in Persia. Similarly, literary representations of Edenic pleasure—like those found in Acrasia's bower of bliss in Edmund Spenser's *The Faerie Queene* or John Milton's Eden in *Paradise Lost*—define pleasure as an erotic, olfactory engagement with nature.

In both Spenser's and Milton's literary gardens, the mutability of pleasure's effects are staged outdoors through scents. It is tempting to conclude after reading Spenser and Milton that sensual olfactory pleasure engenders dangerous sexual appetites; in Milton's *Samson Agonistes,* for example, Samson first experiences Delila through her scent—an "amber scent of odorous perfume her harbinger" like a "stately ship of Tarsus, bound for the isles."[79] Unsure of her sex, he notes only her pefume. The perfumes of King Casper and Cleopatra are both invoked in this image, linking Delila and her sexual trangsressions with her status as Samson's "wife" and "traitress." Perfumes are lavish ornaments, "ornate and gay." However, most who wore perfumes were not conflated with "traiteresses" (l.724, 725). Read alone, garden pleasures always seem hedonistic and dangerous, ignoring the popular appeal of perfumes that actively sought to recreate these spaces indoor. Unlike Spenser's bower of bliss or Milton's Eden, Margaret Cavendish's *Convent of Pleasure* imagines scents as tools of creation rather than loss, using elaborate garden design and potpourri vases as the means of staging erotic pleasure indoors. Released in an erotic, environmentally controlled, women's-only space, Cavendish's scents suggest that the transition from bowers to bedrooms was not always imagined as a loss of pleasure, especially for women. Rather, perfumes were key tools in defining control over their private, and eroticized, domain.

Spenser's bower of bliss is defined by its ephemerality. At the end of Book 2 of *The Faerie Queene,* the two heroic knights, Guyon and King Arthur, famously destroy the garden paradise. The destruction is sudden, violent, and virulent. Acrasia and her lover's erotic bower is "feld," "defaced," and "spoiled." Bliss becomes balefulness; what was once most fair becomes most foul: "But that their blisse he turn'd to balefulnesse: / Their groues he feld, their gardens did deface, Their arbers spoyled, their Cabinets suppresse, / Their banket houses burne, their buildings race, / And of the fairest late, now made the fowlest place."[80] It is hard not to read such destruction as a phenomenological reaction to an encounter with erotic and ambient bliss. On entering Acrasia's garden, Guyon seems mesmerized by the beauty of both Acrasia and her surroundings: "Her cheeks the vermeil red did shew / Like roses in a bed of lilies shed, / The which ambrosiall odours from them threw, / And gazers sence with double pleasure fed, / able to heale the sicke, and to revive the ded." Pleasure here is doubled: the visual beauty of the red roses against the lilies highlights the whiteness of skin, a pleasurable sight amplified by their "ambrosial" odors "thrown" from the flowers, presumably as the lovers gyrate on their bed of roses.

Guyon's violent reaction to Acrasia's floral ecstasy charts a phenomenological overload to the ambient pleasures of the garden. In *Renaissance Self-Fashioning*, Stephen Greenblatt argues that Guyon's seemingly senseless destruction of Acrasia's blissful bower is compelling to contemporary readers because it tests how one approaches the mutability of pleasure, both aesthetic and embodied: "[I]t tests in a remarkably searching way our attitudes toward pleasure, sexuality, the body; tests too our sense of the relation of physical pleasure to the pleasure of aesthetic images and the relation of both of these to what Guyon calls the 'excellence' of man's creation."[81] Guyon's experience of Acrasia's bliss threatens his sense of an autonomous self because it is fused with his aesthetic pleasure in the bower itself, collapsing the beauty of nature with the sexual pleasures imagined to occur there. As Greenblatt expounds, pleasure in the bower of bliss is both a temporal, physical sensation and a cultivated, aesthetic response. Guyon's violence resolves a conflict between these two types of pleasure: Acrasia's prelapsarian garden bliss can not exist in a postlapsarian world.[82] With the destruction of the bower, Guyon asserts that aesthetic pleasure will triumph over inordinate bodily pleasures, at least in the *Faerie Queene*.[83]

Smell is an important part of this process because it leaves the body vulnerable to its environment. The bulk of Book 2 of the *Faerie Queene* emphasizes this point. Prior to his arrival in Acrasia's odoriferous bower of bliss, Guyon witnesses an attack on the Castle of Alma, which literary critic Michael Schoenfeldt terms the "allegorical core of the book."[84] Alma's castle represents the body's vulnerability to external influences: "the self is a castle, but one that . . . must police its necessarily porous borders with a potentially harmful world, winnowing the matter that is allowed in, and expelling any material that could harm the self."[85] In Schoenfeldt's reading of both Alma's and Acrasia's allegorical realms, Alma's castle triumphs over Acrasia's garden because the stomach regulates the body through exhaling vile winds. Juxtaposed with the multisensorial attacks unleashed on the bulwarks of Alma's castle, including smells that would torment even the fiends of hell, Acrasia's bower of bliss emerges as a bodily space of odoriferous—and sexual—bliss.[86] The castle of Alma stages the triumph of bodily balance through its internal systems of control, regulation, exhalation, and excretion (including foul gas). Acrasia's garden celebrates a loss of such control, cultivating instead inordinate (and odoriferous) pleasure in the external beauty (and inhalation) of nature.[87]

Milton's *Paradise Lost* embraces the paradox of pleasure found in Acrasia's garden, exploring Eden's pleasures through the point of view of a character denied access to almost all of them—except smell. Eden is described in the poem through Satan's experience of it. Whereas Guyon is lured by the visual beauty of Acrasia and her bower of bliss, Satan, banished to the sulfuric stink of hell, is beguiled instead by Eden's fragrant, "native" perfumes—perfumes that "whisper whence they stole" of their "balmy soils." Eden's "pure . . . now purer air" is composed of "gentle gales . . . [f]anning their odoriferous wings" that first greet the fiend, who experiences them like a sailor denied sensory variation: "As when to them who sail / Beyond the Cape of Hope, and now are past / Mozambique, off at sea north-east winds blow / Sabean odours from the spicy shore / Of Araby the blest, with such delay . . . / So entertained those odorous sweets the fiend / Who came their bane, though with them better pleased / Than Asmodeus with the fishy fume" (4.153–70).[88]

Eden's pleasures are subjective. Satan's experience of them is marked by his peripheral status: Eden's "native perfumes" contain only painful reminders of his own lost Paradise. The "odoriferous" winds of paradise torture him like the fishy fumes tortured Asmodeus. Likewise, Eden's "native perfumes" are likened to the Sabean odors of the Levant, experienced by sailors navigating past such "spicy" shores, mapping paradise onto seventeenth-century spice routes. Satan's journey to earth is like those sailors traveling south of Africa, around the Cape of Good Hope, past Mozambique toward Yemen, in pursuit of the "Sabean" odors of jasmine and rose.[89] In Dryden's *Aureng-Zebe* (1675), for example, Nourmahal, the Persian empress of Mogul India, has an incestuous dream set in an erotic garden. She dreams about her stepson, Aureng-Zebe, beneath whose feet "were spread / What sweets soe'er Sabean springs disclose, / Our Indian jasmine, or the Syrian rose" (4.100–102). Eden's perfumes mimic those of contemporaneous gardens filled with natural—and artificially enhanced—floral jasmine scents from the east.[90]

Milton's Paradise also matches descriptions of seventeenth-century pleasure gardens. Paradise is imagined as an enclosed space that displays "in a narrow room all of nature's wealth" in order to please and delight "all of human sence[s]," especially smell.[91] There is one key difference: in Eden, odoriferous pleasure need not be cultivated. The tree of life, for example, is planted in an arbor of "trees of noblest kind for sight, smell, [and] taste," but its blooming, ambrosial fruit is the most fragrant (4.241). Here, nature, "not nice art," boons "flowers worthy of Paradise in beds and curious knots" and "Groves whose rich trees wept odorous gums and balm, / Others whose fruits burnished with golden rind" (4.249). And, though Adam is a resident, it is Eve who labors tirelessly to cultivate the garden and define it as her own. Before gardening, she even goes so far as to perfume the space herself, strewing the ground "[w]ith rose and odours from the shrub unfumed" (5.348). Although nature is its own best perfumer, Eve's subtle environmental control over the scents of Paradise links her acts of labor with odoriferous pleasure in Eden, though her desire to cultivate such pleasure leaves her vulnerable to Satan's advances.

The smell of Paradise naturally blends with the aroma of heavenly bliss that arrives on Raphael's angelic wings. Shaking his plumes, Raphael's "heavenly fragrance filled / The circuit wide." Like Satan, Raphael's wings contain environmental traces of his journey to earth: "Glittering tents he passed, and now is come / Into the blissful field, through groves of myrrh, / And flowering odours, cassia, nard and balm; / A wilderness of sweets; for nature here / Wantoned as in her prime, and played at will / Her virgin fancies, pouring forth more sweet / Wild above rule or art; enormous bliss" (5.285–97). The blissful scents of heaven are linked to the "wanton" play of nature in her prime. Generation in this earthly paradise is both erotic and odoriferous. Against this *sprezzatura* of odoriferous generation, Eve and Adam enjoy a sexual bliss unknown to fallen man: "here in close recess / With flowers, garlands, and sweet-smelling herbs / Espoused Eve decked first her nuptial bed" (4.709).

Like Acrasia, Eve's sexual pleasure is amplified by her odoriferous surroundings. Milton's narrator's voice interrupts this glimpse of sexual, and sensuous, paradise: "And on their naked limbs the flowery roof / Showered roses, which the morn repaired. Sleep on / Blest pair; and O yet happiest if ye seek / no happier state and know to know no more" (9.762–74). Eerily echoing Satan's impeding manipulation of Eve's dreams, the narrator's warning gestures toward other contemporaneous uses for floral ingredients in the period, namely soporific and aphrodisiacal effects. Indeed, in her nightmare, Eve imagines an angellike creature offering her a dangerous object with a "pleasant savoury smell" impossible to resist; it "[s]o quickened appetite, that I, methought, Could not but taste" (5.85). When Satan approaches Eve to tempt her with forbidden pleasure, the metaphors of scent and floral bounty are overwhelming. No longer celebrating an innocent sexual bliss, floral scents mark Eve's naiveté. She herself becomes a rose that invites plucking: "Veiled in a cloud of fragrance, where she stood / Half spied, so thick the roses bushing round / About her glowed, oft stooping to support / Each flower of slender stalk, whose head though gay / Carnation, purple, azure, or specked with gold / Hung drooping unsustained, them she upstays / Gently with myrtle band, mindless the while / Herself, though faire unsupported flower / From her best prop so far, and storm so nigh" (9.750). Surrounded by a cloud of fragrant blooms, Eve works to bind and support the flowers, whose heads are too large and colorful for their slender stalks. Her pleasure inspires Satan's: Milton likens her to a "faire unsupported flower," half spied through a cloud of fragrance.

Eve's scent merges with that of the garden, intoxicating Satan and accelerating his plan. Milton's narrator imagines Satan sniffing Eve herself as one of Paradise's sweet floral scents. Again, scents transform into contemporaneous osmologies as they are experienced by Satan: "As one who long in populous city pent, / Where houses thick and sewers annoy the air" (9.445). Linking an urban reader's experience of the city—an experience that echoes Cowely's configuration of London as Sodom—with Satan's experience of Paradise, Milton uses olfaction to demarcate not only prelapserian bliss but also the postlapserian *desire* for floral scents, especially to seventeenth-century readers "pent" in populous cities, crowded with overflowing sewers and nostalgic for rural life.[92]

Satan's experience of Eve's perfume is so powerful that it temporarily distracts him from his own designs, rendering him silent and disarmed. In this moment, odoriferous innocence renders him "stupidly good": "Her graceful innocence, her every air / Of gesture or least action overawed / His malice, and with rapine sweet bereaved / His fierceness of the fierce intent it brought: / That apace the evil one abstracted stood / From his own evil, and for the time remained / Stupidly good, of enmity disarmed / Of guile, of hate, of envy, of revenge" (8.460–65). Eve's scents are "innocent," but they overwhelm Satan with every other emotional response except pleasure. Even in Paradise—when yoked with innocence and bliss—scents have a rapacious power to inspire overwhelming appetites. They disorient him, rendering him temporally and spatially transfixed. For scents also provide a map through such space: the tree of knowledge, as he notes to Eve, is located just "beyond a row of myrtles," near a fountain that blows "of myrrh and balm" (9.626–30). As he recovers, he teaches Eve how to desire such a scent.

In his rhetorical assault on Eve, Satan describes the savory odor of the forbidden fruit as more pleasing than the "smell of sweetest fennel, or the teats / Of ewe or goat dropping the milk at even, / Unsucked of lamb or kid, that tend their / play" (9.579–83). These animal scents are strikingly different from the innocent, floral perfumes of Eden's roses. As in her nightmare, the experience of "savory" scents amplifies and accelerates Eve's desire to satiate her other senses

(9.579). For Adam, too, eating the forbidden fruit immediately inspires a raunchy appetite. Retiring to the shady bed of roses, Adam and Eve enjoy sexual delights until "dewy sleep oppressed them" (9.980). For the first time in Eden, pleasure and bliss are temporally bound. No longer endlessly generative, the lush natural environment of the garden grows overwhelmingly soporific. Pleasure abates and Paradise is lost. With Adam and Eve's exile, a ghastly scent of decay overtakes earth as a "smell of mortal change" descends. So strong is the scent that it wakes the dog of death, who "scented the grim feature, and upturned / His nostril wide into the murky air, / Sagacious of his quarry from so far" (10.270–80). Far from bliss, the entire earth is covered with the scent of disease and decay, of "living carcasses designed for death." In Milton's *Paradise Lost*, like Spenser's bower of bliss, odoriferous pleasures are ephemeral, defined both by their inordinate bliss and by their remarkable loss.

If the garden emerges in early modern culture, to borrow Hawkins's term, as a "goodlie Amphitheater" of multisensorial pleasure, it was more than just a space of loss; it was also a space of paradox, offering pleasures that were both cultivated and inordinate. As both Spenser and Milton strikingly demonstrate, such pleasures were gendered: both Eve and Acrasia are gardeners.[93] The flourishing of early modern gardens occurred at the same time as enclosure acts, allowing private cultivation of nature to emerge alongside powerful political debates about public land use.[94] The increase in private gardening coincided with the disappearance of public lands. Gardening thus appealed to those who were increasingly denied access to more public venues, especially women: "[T]he garden was an escape, a source of renewed vitality, a private domain which the gardener, however beaten down by the world, could order, arrange and manipulate without fear of contradiction . . . this was one of the reasons why gardening so appealed to women, to whom other spheres of activity were closed."[95]

Like the paradox of pleasures afforded in the garden, the most prominent women gardeners in England had a paradoxical power: royal queens and women at court were instrumental to the "gardening revolution" in early-seventeenth-century in England. Henrietta Maria of France, queen consort and wife of Charles I, was particularly important to defining English style, translating French and Italian layouts for English landscapes.[96] Henrietta Maria's formal gardens were known for their striking use of hydraulic waterworks, grottos, and fountains.[97] The queen consort's royal gardens were "goodlie Amphitheater[s]" in more ways than one: these exotic, secluded gardens of pleasure made excellent stages for masques and other aristocratic, theatrical amusements. Ben Jonson's *Love's Welcome to Bolsover* openly celebrated the queen's interests in gardening and her influence on other aristocratic gardens. The masque, performed at Bolsover Castle, the home of William Cavendish, Earl of Newcastle, celebrated Cavendish's newest addition to his garden: "a Garden of Love."[98] The enclosed pleasure garden, lined with jasmine, had an elaborate fountain with statues of Eros and Venus at its center.

One cannot help but wonder if Margaret Cavendish had her husband's "Garden of Love" in mind when she wrote the fanciful *Convent of Pleasure* (1668). Margaret Cavendish most likely never entered the garden at Bolsover Castle; she certainly did not witness Jonson's masques performed there for the queen. Gardens, however, loom large in her play *Convent of Pleasure,* which not only examines property relationships but also uses the metaphor of the pleasure garden to provide a fantastical backdrop for a new staging of pleasure.[99] In it, Lady Happy, a powerful aristocratic woman, creates a women's-only sphere for languishing in multisensorial pleasures, indoors and out.

Lady Happy's overt aim is to create a chaste atmosphere of pleasure for her company of aristocratic women. In her lengthy musings about architecture and landscape architecture, she outlines the sensory pleasures of both the convent and its grounds. Organizing pleasures according to the seasons, Lady Happy creates stunning multisensorial effects. She defines pleasure as a celebration of taste, touch, and smell. She celebrates visual and aural pleasures: there are mirrors in each bedchamber (ostensibly for narcissistic pleasure), rare pictures hung indoors during the summer months, "babbling" streams of scented water next to every bed, and fountains in the garden (2.2). In comparison to these visual and aural stimuli, however, the pleasures of the other senses seem ridiculously ornate. Lady Happy describes precise instructions for luxurious fabrics, elaborate menus, and her garden pleasures.[100] The convent also celebrates natural and artificial scents. Lady Happy orders "[f]loor[s] strew'd with sweet Flowers" in the spring, "strew'd every day with green Rushes or Leaves" in the summer, along with "flowerpots with various flowers." Along with scented streams, each bedchamber is to be hung with "Gilt Leather" scented with "Franchipane" in fall, and finally, she orders "all the lights to be Perfumed Wax" in the winter. Her garden is elaborate, with orange trees and "every season of all sorts of Flowers, sweet Herbs and Fruits" (2.2). Such luxuries shape an environment for women to enjoy worldly sensual pleasures. Self-consciously attempting to create a paradise for women, Lady Happy cultivates her garden and convent to be a chaste, gratifying zone of pleasure. With the arrival of a mysterious and seductive Princess, however, Lady Happy's garden becomes a stage for erotic dalliances. Her convent's perfumes, imagined—albeit temporarily—to be chaste tools of sensuality, ultimately create a very different kind of odoriferous delight.

Feminist scholarship has emphasized the sex-segregation of Lady Happy's "convent" of pleasure.[101] In an article on tactile pleasures in feminine utopias of the Restoration period, Misty Anderson argues that Margaret Cavendish's utopian vision of convents emphasizes women's lack of control over land. The "convent" institutes a "sororal" contract in the place of a sexual contract, which Anderson argues "impedes the transfer of property between men through women," emphasizing instead property "for the immediate pleasure of women."[102] Such a switch, Anderson rightfully points out, has environmental effects. In Cavendish's convent, the "demand for the hetero-normative love plot" is diminished "without foreclosing female pleasure."[103] For Anderson, women's pleasurable relationship to property relies on reclamation of both the body and the space of the home.[104] But, in making her claims, Anderson precludes the possibility that England's gardens were environments under women's control or cultivated for women's pleasure; she provocatively concludes that Cavendish's vision of pleasure is best suited for contemporary media productions. Interpreting the play's preoccupation with multisensorial joys, luxurious materiality, and "fluid" erotic identities, she argues that Cavendish's play asserts a postmodern, feminist logic of pleasure that could not have been performed in seventeenth-century Restoration theater spaces.[105]

For Anderson, Cavendish's emphasis on the materiality of pleasure—the mirrors, fabrics, and waterworks throughout the convent—is imaginative, emphasizing a scopophilic pleasure that seems more familiar to us than to seventeenth-century Restoration drama.[106] Such a reading is compelling, but perhaps Cavendish's text also conceives of pleasure as a series of embodied responses to specific spaces. Lady Happy's convent, like Acrasia's bower of bliss and Eve's Eden, is a cultivated environment designed to amplify the pleasure of *all* the senses, especially olfaction. Cavendish's utopian environment might have been ill-suited for Restoration public theaters, but it was perfectly suited for performances in seventeenth-century gardens of pleasure, like the Garden of Love at Bolsover Castle.

As both a cultivated garden and a luxurious interior space, Lady Happy's "convent" raises provocative questions about women's ability to cultivate odoriferous joys. Using elaborate technology, Lady Happy designs an environment exclusively dedicated to cultivating pleasure. Although she decries the artificiality of many garden features, she employs waters, fountains, and grottos in planning her garden.[107] Outlining her vision to Madam Mediator, Lady Happy delineates her plan to create a "natural" environment where women's "prayers" are spontaneous and "flow from their Mouth" rather than "Spring from their heart, like rain water that runs throw Gutters, or like Water that's forced up a Hill by Artificial Pipes or Cistern" (2.2). She moves fountains indoors and places scented streams of water next to every bed. Her goal is to satiate the senses: "Wee'l please our Sights with Pictures rare; / Our Nostrils with perfumed Air. / Our Ears with sweet melodious sound; Our tast with sweet delicious Meat, / And Savory Sauces we will eat" (1.2). Such multisensorial delights artificially manipulate nature to create an ambient zone of pleasure. As such, Cavendish's play poses important questions about women's happiness and their control of their own pleasure.

Indeed the play begins with three gentlemen querying Lady Happy's future now that her father, Sir Fortune, has died. The gentlemen suggest marriage, but Lady Happy immediately enters the stage and shifts the question. Instead of considering happiness as a result of a fortuitous match, she imagines it as a perpetual effect from a pleasurable environment. As her servant notes, however, Lady Happy embodies happiness because her father has died, leaving her his "fortune." There are economic consequences for such shifts in focus: "Madam, you being young, handsome, rich and virtuous I hope you will not cast away those gifts of Nature, Fortune, and Heaven, upon a Person which cannot merit you?" (1.2). The gentleman and her servants equally worry who will control (and partake in) the lady's happiness.[108]

Lady Happy admits that she has gifts to share but articulates a desire to distribute pleasure to those who need it; in her worldview, unmarried women like herself are sorely in need of pleasure. The pleasures of the "publick" world are far outweighed by its pains—primarily contact with men; the greater a woman's contact with men, the less she experiences pleasure. Imagining her options, Lady Happy is dismayed. Even as the wife of the "best of men," she would still be constrained by her duties, unable to receive pleasing words from potential suitors: "They might gaze on my Beauty, and praise my Wit, and I receive nothing from their eyes, nor lips; for Words vanish as soon as spoken, and Sights are not substantial" (1.2). If she indulges, she imagines herself "turn[ed] Courtizan," but the pleasure gained hardly matches the pain resulting from a loss in health and the inevitable "Quarrels and Brouilleries" that would result with jealous rivals.

Thus, instead of entering the public world of men, Lady Happy chooses to remain in her private home, cultivating her own pleasure. In some sense, this is a chaste redefinition of pleasure, queering virginity.[109] In effect, Lady Happy designs a domestic interior space that heightens "lawful" pleasures, creating not a "Cloister of restraint, but a place for freedom, not to vex the Senses but to please them." Redesigning monastic medieval gardens with a seventeenth-century emphasis on sensory pleasures, Lady Happy announces her intentions: "Thus will in Pleasure's Convent I / Live with delight, and with it die" (1.2).[110] The convent is imagined as a multisensory experience of chaste pleasure, but the dramatic arc of the play queries the effects of such stimuli: how much sensual pleasure can one receive before it becomes sexual?

Lady Happy's garden foregrounds same-sex eroticism, self-consciously exploring the boundaries of "natural" pleasures.[111] As literary critic Valerie Traub argues in the *Renaissance of Lesbianism*, the garden setting is key to the play's negotiation of same-sex desire, enabling the erotic play between Lady Happy and the Princess but also challenging it by defining it within particular pastoral traditions that elegize female homoerotic desire.[112] The garden space amplifies women's pleasure yet precludes its expression in anything other than what is understood to be "natural." For Lady Happy, this means the socially acceptable terms of female friendship. Act four begins with Lady Happy declaring her desire for the Princess: "My Name is Happy, and so was my Condition, before I saw this Princess; but now I am like to be the most unhappy Maid alive: But why may not I love a Woman with the same affection I could a Man? No, no, Nature is Nature, and still will be from all Eternity" (4.1). In Lady Happy's formulation, nature's power is timeless and an uncultivated expression of moral law.

Nevertheless, she and her servants perform a series of masques that detail the horrors of heterosexual unions and increasingly shift toward depicting embraces "of a Femal kind" that gradually grow more "fervent." Celebrating same-sex love in a pastoral setting, these theatrical vignettes stage erotic longing as "natural" but increasingly include songs of "love-sick tales," drinks of "syllabubs," and snacks of "apples drown'd in stronger Ale" (4.1). In her garden of pleasure, sensual pleasures lead to sexual ones: Lady Happy and the Princess's erotic role-playing is both normative and transgressive. Mimicking pastoral poetic traditions and medicinal aphrodisiacs, these vignettes progress toward stronger articulations of sexual desire. Indeed, Madam Mediator worries that Lady Happy's health is deteriorating, intuiting that too much sexual "sport" has weakened her. Immediately after this pronouncement, Lady Happy and the Princess stage a strange theatrical interlude. The Princess, as the Sea-God Neptune, expresses desire for Lady Happy, her Sea-Goddess. In this long articulation of erotic desire, the Princess-Neptune decries the artificial aspects of nature above, imagining her kingdom to be "no base, dissembling, coz'ning Art" in which "Nature directs and doth provide Me all Provisions which I need" (4.1). She ultimately concludes that, in her court, beds are made of ambergris; she is also attended by "mare-men" and "mare-maids" and presented with ships (4.1). Staged in a garden, perhaps next to a fountain whose statuary depicts this exact scene, same-sex erotic pleasures are ironically articulated through "coz'ning" artifice: odoriferous (and ambient) pleasures of the gardens through musk-based scent. The stately, refined, cultivated space of the garden enables an inordinate celebration of same-sex pleasure in nature.[113]

Act five, however, immediately opens with the revelation that the Princess is a man. This dramatic deus-ex-machina precludes the "incloistered" women from expressing erotic "feminine" embraces, reasserting boundaries between "natural" and "artificial" pleasure. As Laura Rosenthal concludes, in the convent, "nature is *not* nature but, like everything else, a performance."[114] Valerie Traub agrees, asking "is Nature, in fact, natural" in the convent?[115] In such interpretations, Lady Happy's "feminine usurpation of Nature" enables new erotic possibilities: her artificial manipulation of the garden cultivates sensual and sexual pleasure. These pleasures are "abandoned, however, when Cavendish exposes the Princess as a Prince."[116] Traub concludes, "Implicit in [Cavendish's] fantasy is the hope that the Prince has learned something of marital responsibilities from his masquerade, and something of women's desires from luxuriating among the convent's many pleasures."[117] In many ways, Cavendish's play offers a disappointing conclusion: Lady Happy's pleasures are redirected, rather awkwardly, into conventional heterosexuality. Her pleasure garden, however, remains. Read within the context of seventeenth-century technologies of gardening and perfuming, as well as the play's attention to staging the environmental influences of pleasure, it is nearly impossible to imagine that the Prince has not "learned" new ways of approaching marital pleasure.

Scents could be a powerful component of sensual and sexual pleasure and were key tools in defining erotic spaces. Olfactory pleasure in nature was one way other kinds of pleasures, especially erotic acts, could be represented. The popular appeal of both pleasure gardens and potpourri provided early modern men and women with tools to cultivate sensual environments designed to elicit certain pleasurable—and erotic—effects, especially through manipulation of the scents of the gardens. The ability of floral scents to "refresh," "exhilarate" and "cheare" the body defined English pleasure gardens and was increasingly linked to erotic pleasure. Similarly, the smoke perfumes emanating from potpourri brought these scents indoors, enabling early modern English men and women to cultivate erotic zones of heightened sensory perception both indoors and out. Whereas we might imagine that the private realm of the bedroom evolved in stark contrast to public acts of sex, as these literary representations suggest, jasmine perfume defined both as ambient zones of pleasure.

I am not the first to suggest that early modern England's bowers were as blissful as its bedrooms.[118] Indeed, many scholars describe early modern England itself as an idyllic sexual zone that reflects a prelapserian, pastoral state of "spontaneous, rather than guilt-ridden" pleasure, or celebrates an "earthy attitude towards sexuality."[119] Such critical metaphors of "earthy" sexuality follow Michel Foucault's famous anecdote about a French farmhand's wandering hands. Foucault's vision of sex outdoors is tinged with nostalgia, a pastoral vision of sexual pleasures imagined to exist on bucolic farms, free from juridical or medical scrutiny that has little to do with the sexual acts that might have occurred in England's garden spaces.[120]

Examining the metaphors of creation and loss that surround both early modern gardens and the pleasures they inspired, I argue that seventeenth-century gardens of pleasure cultivated an olfactory pleasure that came to define certain zones as erotic. Jasmine perfume, in particular, with its strong, unique floral scent, was noted for its ambient, aphrodisiacal qualities. The popularity of both pleasure gardens and potpourri fumigations—both new seventeenth-century phenomena—suggests that the history of bedrooms as sites of sexual pleasure depended on the kind of sensuous, odoriferous pleasure found earlier in English gardens. Potpourri vases filled with jasmine perfumes reflect an attempt to yoke the odoriferous breezes of England's pleasure gardens and release them indoors.[121]

Ephemeral Remains

In his 1751 treatise *Philosophia Botanica*, Carl Linnaeus introduced a radical new system for classifying botanical material that allowed him to attend to both generic and specific qualities of vegetative matter. His system involved surveying a plant's reproductive organs and classifying them through binomial nomenclature; the inclusion of the secondary "diagnostic" term enabled botany to unfold, in his terms, "like territories on a geographical map."[1] Mapping floral reproduction through visual systems of observation, Linnaeus organized all the world's plants into classifiable matter. Given the emphasis on spatial mapping, it is perhaps not surprising that vision emerges as the organizing sense through which all other botanical phenomena are understood; plants are primarily identified through visual concordances between configurations of their reproductive parts.[2] In such a system, smell is almost inconsequential, yet Linnaeus, like so many of his eighteenth-century scientific peers, offers a brief theorization of olfaction before dismissing its importance to scientific study. Though the scent of a plant is an important property (along with its color and its taste) and thus should be included in descriptions, Linnaeus reminds his readers that "scent never clearly distinguishes a species," since the sense of smell is "the most obscure of the senses" and because the "scent of all things very easily varies."[3] Smell also troubles his botanical map, for he notes that "scents do not allow for fixed boundaries and cannot be [spatially] defined."[4] Linnaeus's binomial descriptive system revolutionized European approaches to biological matter; in many ways, it reflects all that Enlightenment science, with its emphasis on quantifiable evidence, teleological progress, and objective truth, would come to represent within modern, Western approaches to intellectual history. It also sounds a historic death knell for olfaction. With the rise of Enlightenment science comes the triumph of the eye over the nose (along with all other sensory organs), or so the story goes.[5] The term *Enlightenment* itself emphasizes such shifts: light was linked with reason and scientific discovery. Eighteenth-century philosophers, influenced by Sir Isaac Newton's paradigm-shifting study of optics in the century before, defined theirs as an age of light.[6] Seventeenth-century scientific discovery thus marks a tri-

umph over the "dark" ages of the medieval period; that leads to further enlightenment, and a long, teleological progress toward modernity begins. The elevation
of vision over all other senses introduced a spatial dimension to reason: a disem-
bodied, universal perspective was valued over more material, embodied ways of
knowing. Olfaction was subjective and thus subjugated. Vision was the domain
of the human, smell, the domain of the animal—so much so that Denis Diderot
could famously "suspect" the blind of inhumanity; for Diderot, lack of vision was
linked to a dangerous lack of morality.[7]

Diderot is not alone in making this claim; countless groups were defined in
such sensory ways. Slaves—both African and native American—were routinely
used by British colonists in the eighteenth century in the American South to collect valuable botanical specimens due to their perceived olfactory acumen, which
was believed to help them navigate the dangerous environmental conditions of
southern swamps.[8] Such sensory insight, though valued, also marked them as
expendable laborers. It offered no paradox for English naturalists to trust the
botanical knowledge generated by such laborers yet denigrate them for it.[9] Likewise, gynecological manuals from the nineteenth century continued to recom-
mend the application of fetid smells to revive "hysterical" women, long after
such practices had been abandoned for men.[10]

These kinds of practices document the shifting social meanings of sagacity
across the historical scope of this book. The term, from the Latin *sagacitas,* linked
acuteness of sensory perception—especially olfaction—with cleverness and in-
telligence. Knowledge was embodied and spatially defined; one was as clever as
one's nose was keen.[11] Ben Jonson's *New Inn* (1629) stages sagacity precisely
opposite to Diderot's conception of it. Olfaction is an index of moral value, punning on the link between smell and wisdom: in the play, a character wonders
aloud how the host of the titular inn, a man of such "sagacity, and clear nostril,"
could work in such a sordid industry (1.3.110).[12] Having a "clear" nostril was a
positive quality, at times even reflecting a keen aptitude for regal leadership,
particularly if one could smell deception or treason in tumultuous times. Edward Hall's *Union of the Two Noble and Illustre Famelies of Lancastre [and] Yorke*
describes how Henry VII, in the second year of his reign, discovered a treason-
ous conspiracy against him that smelled of "a dunghill knave and evil born vil-
lainy."[13] Hall also chronicles how, in the third year of his reign, King Henry sent
two ambassadors to France, both renown for "prudent sagacity" as well as their
"fatherly and wyse personage." As such, they were particularly adept at negotiating the tense political climate between France and England.[14] Likewise, James I
was repeatedly described in contemporaneous and later historical accounts as

sagacious: his ability to detect the gunpowder in the Gunpowder Plot turned on his status as an "excellent smeller."[15]

By the end of the seventeenth century, however, *sagacity* was beginning to connote a dangerously embodied epistemology, often invoked to explain religious and ethnic others. In 1679, the "controversialist" Lancelot Addison, for example, describes "the sagacity of this great Cheat Muhammed" to explain the influential power of his prophecies.[16] By the start of the eighteenth century, the layered meanings of sagacity further unraveled. *Sagacity,* as a term, described intuitive or instinctual knowledge, the very opposite of rationality.[17] Its association with sensory perception, specifically smell, described brute instinct, usually applied to animals. The senses of taste, touch, and smell, defined as proximate and intimate, thus remained mired in a dangerous bodily materiality, linking its meaning to distinct times and spaces rather than a universal ideal.

In hindsight, it is easy to see how such an approach connects to the horrors of post-Enlightenment history: with perspective comes a privileging of distance and a wariness of physical proximity to others. Our bodies became discrete entities rather than a shared platform of collective environmental conditions. The association of olfaction with unmediated perception, seemingly offering a temporary respite from such a bounded sense of ourselves, also threatens to unravel subjectivity, so much so that Max Horkheimer and Theodor Adorno, in their postmodern *Dialectic of Enlightenment,* can claim that olfaction, more than any other sensory approach, contains an unmediated desire to "lose oneself in the Other."[18] Furthermore, smell—"as both perception and perceived"—provides for a loss of selfhood, which contains a sensuous, primordial "nostalgia for what is lower . . . a longing for immediate union with surrounding nature, with earth and slime."[19]

As I argued at the start of this book, we should be skeptical of such large-scale claims about olfaction and its ability to connect us to the past (and, it should be said, of any "unmediated" desire we might have to smell "the Other"). Such a claim quietly insists on the temporal backwardness of smell within post-Enlightenment culture, defining the past as malodorous and the present as deodorized, even as it seeks to bridge such temporal gaps.[20] Such overly broad claims about the demise of olfaction in post-Enlightenment Western culture obscure the ways such sensory shifts were experienced, embodied, even perceived, by men and women in the past.

Though there is much more to say about such sensory shifts and their influence on aesthetic, economic, political, technological, and intellectual approaches to modernity, I end with Linnaeus in order to excavate what remains buried within his influential treatise and to muse briefly on how such a methodological approach might contribute to larger inquiries into the histories of the body, of its materiality, and of sensation. I do so in order to trouble clear distinctions between pre-Enlightenment modes of perception and post-Enlightenment ways of approaching the material world. Though Linnaeus dismisses olfaction as a tool for taxonomic study, his characterization of it offers a useful reminder of the main tenets of my argument. He argues that the ephemeral materiality of smell challenges any attempt to define or fix it in metaphoric or material space. Despite this assessment, he proffers that all of the world's plants emit one of five odors: ambrosial (such as amber and musk), fragrant (like jasmine), spicy (like sassafras, camphor, and citrus fruit), noisome vegetative scents (like cannabis or opium), or nauseous vegetative scents (like tobacco).[21] He thus offers an archive of scent very similar to mine, even as his binomial taxonomy erases its significance to scientific study.

Linnaeus does not dwell on the tension between olfactory materiality and its descriptive categories; his is a forward-looking project. The distance between the two looms larger for those of us interested in historicizing olfaction. That smells are worthy of scientific or historical investigation is premised on the fact that they materially exist, even though they cannot be seen. Ironically, this premise was established through seventeenth-century investigations of biological matter and its boundaries, investigations linked to the demise of olfaction because they sought to trouble subjective sensory ways of knowing revealed through descriptive language. Such investigations merely redefined the invisible power of air from the scent of divinity, royalty, foreignness, disease, luxury, or nature to that of pathogens, microbes, and molecules. Thus, a productive friction emerges from an attempt to combine material approaches to the body with discursive approaches to perception. Literary evidence like descriptions of Richard Wyche's perfumed saintly remains, Henry VIII's rose-scented performances of kingship, John White's anxious sniffs of sassafras, Londoners' rosemary-scented encounters with plague, Lady Davies's ambergreased angel, and the odiferous (and erotic) pleasure cultivated in Spenser's bower of bliss, to name just a few examples from this book, all gesture toward the same profound materiality that intrigued Enlightenment scientists: invisible, yet perceptible, smells.

The ephemeral history of perfume I have charted abuts this period of scientific investigation into the materiality of olfaction that is often cited as part of its cultural denigration. Its history thus shapes, and was shaped by, these broader investigations into the interrelatedness of bodily perception, materiality, and environment. In the space that remains, I wish to consider how the historical paradoxes of both scents (as material objects) and smell (as a biological and cultural process) provide useful reminders of what we can—and cannot—understand about the past.

As both Linnaeus and I have argued, olfaction troubles the boundaries between bodies and environments. A history of smell is thus a history of the air itself, a history, Michel Serres argues, that is a "vector of everything" yet is also at the "very limit of the insensible."[22] It is composed of confused encounters between things, beings, knowledge, memory, space, and time.[23] This is not a new observation. As early as 1620, Francis Bacon argued that air itself should be historicized.[24] Yet air, like the ephemeral scents that made it perceptible, troubled categorization; as John Donne put it, a scented vapor was a "near-nothing," resisting, as Linnaeus would argue a century later, temporal and spatial containment.[25] Efficacious, yet invisible; powerful, yet miniscule; influential, yet ahistorical— scented air remained a paradox in the seventeenth century. By the end of the seventeenth century, however, Robert Boyle would answer Bacon's call, and in doing so, seemingly pull apart these two terms. In his *Experimental Discourse of Some Unheeded Causes of the Insalubrity and Salubrity of the Air* (1690), he notes that perfume, despite its strong smell, lost almost none of its substance over time.[26] Such research was influential in his work on the inverse relationship between pressure and volume of air (known as Boyle's Law), defining musk as dangerously stable while air was vulnerable and ephemeral. Perfume could pollute the air; so, too, could bodily odors, especially malodorous breath and effluvia.

As Alain Corbin argues in his influential history *The Foul and the Fragrant: Odor and the French Social Imaginary*, eighteenth-century research focused on links between olfaction, public health, and air quality: research like Antoine Lavoisier's work on the smell of prisons (and the rise of carbonic gas in heavily populated areas), posited that respiration was a process of metabolic combustion. Crowded areas—like prisons or cities—were especially dangerous zones. Alexander von Humboldt and Joseph Louis Gay-Lussac soon connected Lavoisier's work on metabolic combustion to decreased oxygen content in such crowded spaces; the body was vulnerable in such spaces but also left its mark, a gaseous residue that polluted the zone for later inhabitants, compounding the dangers latent in populous urban spaces like London and Paris. Louis Pasteur's theory of microbes in the nineteenth century shifted the focus from the smell of purification to the spread of disease by other kinds of "invisible" matter.[27]

By the nineteenth century, perfume is a footnote within a broader "history" of air, but it was key to the emergence of air as a scientific category of analysis, which bears the traces of its influence. From a certain vantage point, postmodern anxieties about invisible, scented air and its relationship to health have come full circle, mapping neatly onto early modern ones. Health advocates, for example, are currently targeting perfume industries for their reliance on phthalate and xeno-estrogenic compounds, endocrine disruptors that cause hormonal imbalances while allowing us to perfume our homes and bodies with the scent of (synthesized) musk.[28] But from another angle, the fact that the scent of musk is once again dangerous obscures the ways in which the scent itself has changed; synthesized musk bears almost no relationship to essential oils derived from the adult Siberian male deer, *moschus mochisferus*. One smells powdery; the other, like an adult male deer.

Clearly that fact matters, even within an ephemeral history of perfume. Because my emphasis in these chapters has focused first on establishing material practices and then on arguing how and why these scents mattered in particular early modern locations, there is a danger that I have inadvertently fetishized the material history of scents in order to approach more nebulous cultural meanings of early modern olfaction. I do not wish to suggest, for example, that re-creating "authentic" early English rosewater or perfumed gloves, for example, from manuscript recipes, or igniting the few English censers that are locked away in museum cases would provide a better understanding of what sixteenth-century roses, ambergris, or incense smelled like to sixteenth-century men and women. Constructing Lady Happy's garden of pleasure gets us no closer to understanding the olfactory pleasure cultivated in early modern gardens. Nor does hunting for sassafras trees in North Carolina reveal more information about John White's encounter with new world ecologies. The world has changed and we along with it, musks and all.

This is not to say that we should dismiss our desire for material evidence—to encounter the past—for it, too, bears an important, ephemeral trace of meaning that is worthy of investigation, particularly since it might help us understand olfaction's role—not in the past, but in the present. This awareness shapes the innovative approach to the collection in the Osmothèque, a museum of perfumery in Versailles, France. Dedicated to archiving both modern perfumes and recreating *parfums disparu* ("disappeared" or "lost" perfumes), the Osmothèque provides a way to phenomenologically experience perfumes of the present and the past simultaneously. Juxtaposing commercial perfumes—and their history—alongside their ephemeral historical antecedents, the museum argues that perfume is an enigmatic, shifting cultural object produced from a wide variety of cultures, periods, and ingredients. Willing participants can sniff these re-created or archived specimens along with the curators and learn more about these distinctions.

Yet these disparate scents are linked by their fragility. The museum defines perfume as an ephemeral object; as such, it does not matter if that ephemerality occurred in the past, is occurring in the present, or will occur in the future. Rather than rely on transhistorical claims about olfaction, the museum instead offers a thoughtful meditation on ephemeral materiality. This approach, no doubt, stems from practicality: perfume fades. Thus, perfume samples are housed in a climate-controlled basement so that their shelf life might be extended. An inert gas temporarily halts oxidation. Although these specimens are preserved so that we might encounter these ephemeral materials of the past, each sniff renders it more vulnerable to oxidation. From a materialist point of view, the archive can only retard an inevitable process.

Its focus on *les parfums disparu* is thus both historical and presentist. Besides its historical re-creations, the museum collects contemporary perfumes, scents that will one day fade from memory. These include popular commercial varieties along with others that are less so. For example, the collection has a substantial number of perfumes released in the last decade of the USSR, scents of poor quality that are not generally recognized as important in the history of perfume, per se, but have important cultural relevance for those seeking to understand the lost sensory worlds of communism. The same might also be said for the lost sensory worlds of capitalism: archived alongside those communist perfumes was the first perfume I ever bought for myself, an artifact from suburban mall culture of New Jersey in the 1980s.

Thus, my archive of scent—along with Linnaeus's and the Osmotheque's— reveals that embodiment is an ongoing epistemological process. Our bodies and their boundaries are shifting contested zones of influence and bear the trace of our complicated histories, both personal and collective. The irony, of course, is that olfaction, as a theory of bodily operation attuned to processing invisible materiality, may enable us to "see" this trace more clearly. Smell can provide fascinating insights into the relationship between material objects, the body, and embodiment, as long as we are aware of the ways it shapes our analytical approaches. To perfume the past is a very different process from uncovering the ephemeral history of perfume. This is immediately clear if one switches from cultural memory toward more intimate and private encounters. As philosopher Michel Serres provocatively argues in *The Five Senses,* "[A]s far as the nose is concerned, the emanations of whomever we have loved remain. It returns to haunt our skin, at dawn on certain mornings. Love perfumes our lives, aromas resurrect encounters in all their slendour . . . life itself announces its presence from afar with these balmy encounters."[29] Such scents, though lost, hover on the borders of embodied knowledge and collective memory, connecting across bodies, space, and time, filled with the promise of imaginative resurrections and with the ephemerality of life itself.

Introduction · Strange, Invisible Perfumes

1. "Press Release: 2004 Nobel Prize in Physiology or Medicine," Nobel Assembly of the Karolinska Institute in Stockholm, October 4, 2004, http://nobelprize.org/medicine/laureates/2004/press.html (cited January 5, 2011).

2. Lawrence K. Altman, "2 Americans Win Nobel for Demystifying Sense of Smell," *New York Times*, October 4, 2004.

3. "Press Release: 2004 Nobel Prize."

4. Axel and Buck use the term *logic* to describe their work. Buck's work extended scientific understanding of how the nose genetically recognizes smells; Axel's work mapped the parts of the brain employed to discriminate between them. Both have designed lectures, available online, for the general public explaining their scientific endeavors: see Richard Axel, "Scents and Sensibility: Towards a Molecular Logic of Perception," Columbia University, New York, NY, May 13, 2004, available at http://c250.columbia.edu/c250_events/symposia/brain_mind/brain_mind_vid_archive.html (cited January 5, 2011); Linda B. Buck, "The Logic of Smell," Karolinska Institute, Stockholm, Sweden, December 7, 2001, available at http://nobelprize.org/nobel_prizes/medicine/laureates/2004/buck-symp.html (cited January 5, 2011). There is still debate over whether the olfactory bulb recognizes smell molecules through shape or vibration; see Rachel Hertz, *The Scent of Desire: Discovering Our Enigmatic Sense of Smell* (New York: HarperCollins, 2008), 27.

5. The Ig Nobel award is a spoof of the Nobel prize, giving awards to serious, but bizarre or funny, scientific studies. The award is designed for studies that "first make people laugh, and then make people think." See "Improbable Research: The Ig Nobel Prizes," http://improbable.com/ig/ (cited January 5, 2011).

6. Ed Frauenheim, "Nothing Fishy about the Sweet Smell of Nobel Success," *New York Times*, October 6, 2004.

7. Ibid.

8. Guatam Naik, "The Smell of Success Isn't Always Sweet in This Line of Work: Mr. Knight Brews Up Odors of Sweaty Locker Rooms and Egyptian Mummies," *Wall Street Journal*, October 11, 2004.

9. Axel, "Scents and Sensibility."

10. When one smells something, genetically programmed smell receptors dispersed throughout the nose receive particular odor-molecules. In order to recognize the smell as a "rose," the brain simultaneous processes numerous molecules. For this reason, odors are usefully thought of as "chords." The process of identifying molecules A, B, and C as a combination that translates into the smell "rose" is termed *binding*. Once this occurs, it is much faster to identify the smell of "rose" versus lilac, lily, or pine scents.

11. For example, as one medical text notes, there are very few cross-cultural studies on the science of smell. "If, as is generally acknowledged, a gap exists in our understanding of olfaction, then with respect to understanding geographic, cultural, and individual differences in olfaction, the hole is a virtual chasm"; Thomas V. Getchell, *Smell and Taste in Health and Disease* (New York: Raven Press, 1991), 287. See also Wendy Smith and Clair Murphy, "Epidemiological Studies of Smell: Discussion and Perspectives," in *International Symposium on Olfaction and Taste,* ed. Thomas E. Finger, Annals of the New York Academy of Sciences, vol. 1170 (Boston: Blackwell, 2009), 569–73, 569.

12. Bruce Smith, *The Acoustic World of Early Modern England: Attending to the O-Factor* (Chicago: University of Chicago Press, 1999); Gina Bloom, *Voice in Motion: Staging Gender, Shaping Sound in Early Modern England* (Philadelphia: University of Pennsylvania Press, 2007); Keith Botelho, *Renaissance Earwitnesses: Rumor and Early Modern Masculinity* (Basingstoke, U.K.: Palgrave Macmillan, 2009); Elizabeth Harvey, *Sensible Flesh: On Touch in Early Modern Culture* (Philadelphia: University of Pennsylvania Press, 2003); Katherine Craik, *Reading Sensations in Early Modern England* (New York: Palgrave Macmillan, 2007); Holly Dugan "Scent of a Woman: Performing the Politics of Smell in Late Medieval and Early Modern England," *The Journal of Medieval and Early Modern Studies* 38.2 (2008): 229–52; Jonathan Gil Harris "The Smell of Macbeth," *Shakespeare Quarterly* 58.4 (2007): 465–86; Peter Charles Hoffer, *Sensory Worlds of Early America* (Baltimore, MD: Johns Hopkins University Press, 2004); Jeffrey Masten, "Toward a Queer Address: The Taste of Letters and Early Modern Male Friendship," *GLQ: A Journal of Lesbian and Gay Studies* 10.3 (2004): 367–384; Marcy Norton, *Sacred Gifts, Profane Pleasures: The History of Tobacco and Chocolate in the Atlantic World* (Ithaca: Cornell University Press, 2010).

13. The work that has been done is very strong. For example, Alain Corbin's landmark study of foul and fragrant smells in eighteenth-century France demonstrated provocatively that olfaction could be historicized; *The Foul and the Fragrant: Odor and the French Social Imagination* (Cambridge, MA: Harvard University Press, 1988). Constance Classen's work on the anthropology of olfaction (and the history of the rose) demonstrates how studying olfaction in the past can illuminate its role in the present; *Worlds of Sense* (New York: Routledge, 1993). Likewise, Susan Harvey's work on early Christian scenting practices; Janice Carlisle's work on smell, class, and masculinity in Victorian literature; and Connie Y. Chiang's analysis of the role of olfaction in shaping American racism at the turn of the twentieth century demonstrate how the study of olfaction can intersect with other research inquiries. See Susan Ashbrook Harvey, *Scenting Salvation: Ancient Christianity and the Olfactory Imagination* (Berkeley & Los Angeles: University of California Press, 2006); Janice Carlisle, *Common Scents: Comparative Encounters in High-Victorian Fiction* (New York: Oxford University Press, 2004); and Connie Y. Chiang, "Monterey-by-

the-Smell: Odors and Social Conflict on the California Coastline," *Pacific Historical Review* 73.2 (May 2004): 183–214. The work on early modern olfaction is equally suggestive. For example, Mark S. Jenner's influential article on deodorization in early modern England demonstrated how the meaning of certain smells (like garlic) depended on their environment; Julian Yates's work on early modern toilets analyzed olfaction's role in the development of privacy in the early modern period; and Jonathan Gil Harris's recent work on the smell and the performance history of *Macbeth* argued that perfumes were important stage properties. See Mark S. Jenner, "Civilization and Deodorization? Smell in Early Modern English Culture," in *Civil Histories: Essays Presented to Sir Keith Thomas,* ed. Peter Burke, Brian Harrison, and Paul Slack (New York: Oxford University Press, 2000), 127–44; Julian Yates, *Error, Misuse, Failure: Object Lessons from the English Renaissance* (Minneapolis: University of Minnesota Press, 2003); and Harris, "Smell of Macbeth." These examples demonstrate the importance of studying olfaction in the past.

14. C. M. Woolgar, *The Senses in Late Medieval England* (New Haven, CT: Yale University Press, 2006), 117.

15. Ibid.

16. Arjun Appadurai summarized the relationship between words and things as one central to anthropologies and histories of commodities. "Commonsensical" oppositions between words and things, Appadurai reminds us, are built on Enlightenment separations between material objects and sensory perceptions; "Introduction: Commodities and the Politics of Value," in *The Social Life of Things: Commodities in Cultural Perspective* (New York: Cambridge University Press, 1986), 3.

17. George Lakoff and Mark Johnson, *Metaphors We Live By* (Chicago: University of Chicago Press, 1980), 239.

18. Susan Stewart, *Poetry and the Fate of the Senses* (Chicago: University of Chicago Press, 2002), 15.

19. Richard Strier and Carla Mazzio, "Two Responses to 'Shakespeare and Embodiment: An E-Conversation,'" *Literature Compass* 3.1 (March 2006): 15–31.

20. Franco Moretti, *Graphs, Maps, Trees: Abstract Models for a Literary History* (New York: Verso, 2005), 2.

21. In making this claim, I align myself with Dominique Laporte's argument in *History of Shit,* which analyzes the "semantic atrophy" of olfactory language in English in the seventeenth century; *History of Shit,* trans. Nadia Benabid and Rodolphe El-Khoury (Cambridge, MA: MIT Press, 2002), ix.

22. See Mark M. Smith, *How Race Is Made: Slavery, Segregation, and the Senses* (Chapel Hill: University of North Carolina Press, 2006).

23. Renaissance technologies of print issued forth new transactions between "material language and the material bodies of readers and writers." Such transactions, Katherine Craik argues, produced new kinds of bodily and literary pleasure; *Reading Sensations,* 3. This pleasure was both physical and cultural: early modern writing practices were overtly visual, but they were also tactile, and, as Jeffrey Masten's and Kim Hall's work on sugar suggest, perhaps even gustatory. See Kim Hall, "Culinary Spaces, Colonial Spaces: The Gendering of Sugar in the Seventeenth Century," *Feminist Readings of Early Modern Culture: Emerging Subjects,* ed. Valerie Traub, M. Lindsay Kaplan, and Dympna Callaghan

(New York: Cambridge University Press, 1996), 168–90; Masten, "Toward a Queer Address." Speech, while overtly aural, was also olfactory, as Coriolanus's horror at the rank breath of the plebeians demonstrates; William Shakespeare, "Coriolanus," *The Norton Shakespeare: Based on the Oxford Edition,* 2nd ed., ed. Stephen Greenblatt, Walter Cohen, Jean E. Howard, and Katharine Eisaman Maus (New York: W.W. Norton, 2008), 3.3:124–27. And, as Carla Mazzio has argued, the shift toward silent reading practices (and perhaps toward more metaphorical rather than material sensations) reconfigured representations of love through technologies of print culture rather than writing cultures. See Carla Mazzio, "Acting with Tact: Touch and Theater in the Renaissance," in *Sensible Flesh: On Early Modern Culture,* ed. Elizabeth Harvey (Philadelphia: University of Pennsylvania Press, 2003), 159–86. These technologies certainly produced new pleasures, but they also produced new kinds of bodily pain. The labor to produce the heft of a book, particularly if it was heavily bound, armored, or locked, reoriented the body around the experience of pain in much the same way as reading practices reoriented the expression of pleasure. Early modern reading practices could provide a useful schema for approaching multisensorial sensation.

24. See, for example, Marshall McLuhan, *The Gutenberg Galaxy: The Making of Typographic Man* (Toronto: University of Toronto Press, 1962), 138. See also Mark M. Smith's discussion of the "great divide" in *Sensing the Past: Seeing, Hearing, Smelling, Tasting, and Touching in History* (Berkeley & Los Angeles: University of California Press, 2008), 8–11.

25. Three histories of olfaction are often cited: Diane Ackerman, *A Natural History of the Senses* (New York: Random House, 1990); Constance Classen, David Howes, and Anthony Synnot, *Aroma: The Cultural History of Smell* (New York: Routledge, 1994); and Annick Le Guérer, *Scent: The Mysterious and Essential Powers of Smell,* trans. Richard Miller (New York: Turtle Bay Books, 1992). In these studies, olfaction—as a bodily process and as a shared cultural phenomenon—is theorized in general terms. Seeking to "demonstrate the importance of odour and olfactory codes in both Western and non-Western societies," and to "bring smell out of the Western scholarly and cultural unconscious into the open air of social and intellectual discourse," the methodological imperatives of these histories are "comprehensive, not exhaustive"; Classen, Howes, and Synotte, *Aroma,* 9–10. These histories argue that olfaction should be scrutinized in greater detail: surveying a wide variety of cultural and historical locations, they seek to underscore the role of the senses in understanding modern embodiment. In *Aroma* this general view enables a broad, cross-cultural approach to the biological and discursive histories of olfaction; whereas in Le Guérer and Ackerman's explorations, olfaction is defined as an "essential, mysterious process" and a "natural" phenomenon. The stability of the body obscures the social and historical networks of power that fueled olfaction's discursive meanings.

26. Edmund Husserl, "A Phenomenology of Reason," in *General Introduction to a Pure Phenomenology* (The Hague: M. Nijhoff, 1982), 231.

27. Recent scholarship in literary studies, anthropology, and history explicates the relationships between the other senses and materiality in a wide range of historical periods. These works are discussed later in this introduction.

28. For a summary of the state of the field of sensory history, see Smith, *Sensing the Past*. For a summary of how this field has impacted early modern studies, see Holly Dugan, "Shakespeare and the Senses," *Literature Compass* 6.3 (2009): 726–40, and Patricia Cahill, "Take Five: Renaissance Literature and the Study of the Senses," *Literature Compass* 6.5 (2009): 1014–30.

29. Mark M. Smith, "Producing Sense, Consuming Sense, Making Sense: Perils and Prospects of Sensory History," *Journal of Social History* 40.4 (Summer 2007): 841–58, 841.

30. George H. Roeder Jr., "Coming to Our Senses," *Journal of American History* 81.3 (1994): 1112–22; David Howes, "Forming Perceptions," in *Empire of the Senses: The Sensual Culture Reader,* ed. David Howes (Oxford: Berg, 2005), 399–402.

31. Bloom, *Voice in Motion;* Dugan, "Scent of a Woman"; Harris, "Smell of Macbeth"; Harvey, *Sensible Flesh;* Hoffer, *Sensory Worlds;* Masten, "Toward a Queer Address"; Norton, *Sacred Gifts;* Smith, *Acoustic World.*

32. Kathryn Linn Guerts, *Culture and the Senses: Bodily Ways of Knowing in an African Community* (Berkeley & Los Angeles: University of California Press, 2003), 8.

33. Michael Schoenfeldt, *Bodies and Selves in Early Modern England: Physiology and Inwardness in Spenser, Shakespeare, Herbert, and Milton* (New York: Cambridge University Press, 1999), 6.

34. Thomas Laqueur, *Making Sex: Body and Gender from the Greeks to Freud* (Cambridge, MA: Harvard University Press, 1990), 15.

35. Judith Butler, *Bodies That Matter: On the Discursive Limits of Sex* (New York: Routledge, 1993), 35. Literary critic Valerie Traub raises a similar point but notes the importance of historicizing the body, arguing that synechdochal representations of the body within psychoanalytic and medical traditions reify the presumption that there is "in fact a whole to be represented." For her influential discussion of how metaphors of the body circumscribe the potential for desire and limit sexual practices for women in the early modern period, see Valerie Traub, "Psychomorphology of the Clitoris," *GLQ: Journal of Gay and Lesbian Studies* 2.1–2 (1995): 107–93, and *The Renaissance of Lesbianism in Early Modern England* (New York: Cambridge University Press, 2002), 226.

36. Hoffer, *Sensory Worlds,* 2; Peter Stallybrass and Allon White, *The Politics and Poetics of Transgression* (Ithaca, NY: Cornell University Press, 1986), 192.

37. Hoffer, *Sensory Worlds,* 2.

38. Consider, for example, James Axtell's metaphor of vivification for the writing of history. Employing the body as a metaphor for narratives, Axtell writes: "The first task in writing history is to reanimate the known facts, which come lifeless from the page. We must *revivify, resurrect,* and *re-create* the past for ourselves." Axtell's image of the "revivified" facts of history is an attempt to transform the "dead" facts of the past into living, breathing beings; *Beyond 1492: Encounters in Colonial North America* (New York: Oxford University Press, 1992), 10.

39. See Katherine Craik, *Reading Sensations,* 3; Mary Douglas, *Purity and Danger: An Analysis of the Concepts of Pollution and Taboo* (New York: Routledge, 1966), 4; Elizabeth D. Harvey, "Introduction: The Sense of All Senses," in *Sensible Flesh: On Touch in Early*

Modern Culture (Philadelphia: University of Pennsylvania Press, 2003), 15; Gail Kern Paster, *The Body Embarrassed: Drama and the Disciplines of Shame in Early Modern England* (Ithaca, NY: Cornell University Press, 1993), 7; Schoenfeldt, *Bodies and Selves*, 2; Smith, *Acoustic World*, 21.

40. Kathleen Canning, "The Body as Method? Reflections on the Place of the Body in Gender History," *Gender & History* 11.3 (1999): 499–513, 499.

41. Ibid., 504.

42. Ibid., 505.

43. For a critique of objectivity in historical methodologies, see Joan Wallach Scott, *Gender and the Politics of History: Gender and Culture* (New York: Columbia University Press, 1988). In philosophy, see Susan Bordo, *Unbearable Weight: Feminism, Western Culture, and the Body* (Berkeley & Los Angeles: University of California Press, 2004), 227.

44. Hoffer, *Sensory Worlds*, 6.

45. Ibid., 11–15.

46. Mark M. Smith identifies this "quiet ahistorical epistemological tendency" as a key component of an emerging split in the field of sensory history. See Smith, *Sensing the Past*, 118.

47. Most "musk" used in commercial modern perfumer is synthetic; its smell is powdery rather than animal.

48. Virginia E. Richardson and Amanda Smith Barusch, *Gerontological Practice for the Twenty-First Century* (New York: Columbia University Press, 2006), 24. Those suffering from anosmia usually report a diminished sense of taste. Keith L. Moore, Arthur F. Dalley, A.M.R. Agur, *Clinically Oriented Anatomy* (New York: Lippincott, Williams & Wilkins, 2006), 1130.

49. Hoffer, *Sensory Worlds*, 20.

50. Smith, *Sensing the Past*, 5.

51. See Richard Palmer, "In Bad Odour: Smell and Its Significance in Medicine from Antiquity to the Seventeenth Century," in *Medicine and the Five Senses*, ed. William F. Bynum and Roy Porter (New York: Cambridge University Press, 1993), 61–68, 62.

52. Plato, *Timaeus*, trans. Francis MacDonald Cornford (New York: Routledge, 1996), 66d–67a, cited in Harvey, *Scenting Salvation*, 103.

53. See Harvey, *Scenting Salvation*, 103.

54. Ibid., 31.

55. Helkiah Crooke, *Mikrokosmographia. A Description of the Body of Man* (London, 1615), 617–19. For Crooke, the nose is the vehicle that draws air into the body while the mind interprets the odor: "The body therefore that perceiueth or apprehendeth odours is placed higher . . . It remaineth therefore that it must be those processes which because they are somewhat like the nipples of a Dugge are called Mammillares" (619). Olfaction blurs the line between the "perceiving body" of man and his more animal instincts.

56. Ibid., 713.

57. Ibid.

58. Ibid., 648.

59. Ibid., 705. Modern scientists preserve this concept: smell is the only unmediated sense, with external stimuli directly interacting with the brain's receptors in the prefron-

tal cortex rather than relaying through the thalamus. See Warrick Brewer, David Castle, Christos Pantelis, and Peter Doherty, *Olfaction and the Brain* (New York: Cambridge University Press, 2006), 15.

60. See Jenner, "Civilization and Deodorization," 136.

61. Stephen Batman, *Batman uppon Bartholome* (London, 1582), sig. C v.

62. Robert Burton, *The Anatomy of Melancholy* (Oxford, 1621), sig. C v.

63. As Ambroise Paré summarizes, "Smelling (according to *Galens* opinion) is performed in the Mamillary processes produced from the proper substance of the braine, and seated in the upper part of the nose: although others had rather smelling should be made in the very foremost ventricles of the braine"; *The Workes of That Famous Chirurgion Ambrose Parey*, trans. Th. Johnson (London, 1634), sig. C6 v, 24.

64. Guy de Chauliac, *The Questyonary of Cyrurgynes with the Formulary of Lytll Guydo in Cyrurgie* (London, 1542), sig. C4 r.

65. Anonymous, *The Problemes of Aristotle* (Edinburgh, 1595), sig. B4 v. On early modern London's "sinks," see Holly Dugan, "Coriolanus and the 'Rank-scented Meinie,'" in *Masculinity and the Metropolis of Vice: London, 1550–1750*, ed. Amanda Bailey and Roze Hentschell (New York: Palgrave, 2010): 139–59.

66. Lanfranco, *A Most Excellent and Learned Woorke of Chirurgerie* (London, 1565), sig. Cgiii r–v.

67. Thomas Tomkis, *Lingua; or the Combat of the Tongue, and the fiue senses for superiority* (London, 1607), sig. E3 v.

68. Ibid., sig. I r.

69. J. H., *A Holy Oyl; and, a Sweet Perfume: Taken out of the Sanctuary of the Most Sacred Scriptures . . . Poured Forth into Eight [or Rather Five] Vessels* (London, 1609), sig. P v.

70. Ibid., sig. Hhh2 r.

71. Jean Howard, "Scripts and/versus Playhouses: Ideological Production and the Renaissance Publish Stage," in *The Matter of Difference: Materialist Criticism of Shakespeare*, ed. Valerie Wayne (Ithaca, NY: Cornell University Press, 1991), 226.

72. Pierre Bourdieu, *Outline of a Theory of Practice* (New York: Cambridge University Press, 1977), 94.

73. As historian Mark M. Smith argues, building on Paul Gilroy's work on race-based thinking, "the human sensate has had to be educated to the appreciation of racial difference"; Smith, *How Race is Made*, 10; Paul Gilroy, *Against Race: Imagining Political Culture Beyond the Color Line* (Cambridge, MA: Belknap Press of Harvard University Press, 2000), 7, 8, 242.

74. See Bloom, *Voice in Motion*; Carlisle, *Common Scents*; Alain Corbin, *The Foul and the Fragrant: Odor and the French Social Imagination* (Cambridge, MA: Harvard University Press, 1986), Martin Jay, *Downcast Eyes: The Denigration of Vision in Twentieth-Century French Thought* (Berkeley & Los Angeles: University of California Press, 1994); Smith, *Acoustic World*.

75. Leslie Adelson, *Making Bodies, Making History: Feminism and German Identity* (Lincoln: University of Nebraska Press, 1993), xiii, quoted in Canning, "Body as Method," 505.

76. The phrase "spatial practice" derives from Henri Lefèbvre, *The Production of Space* (Oxford: Blackwell, 1991). In this study of the historical production of social space, Lefèbvre theorized that accessing the social-ness of space means exploring its production through both artistic representations and cultural norms of the past: "If space is produced, if there is a productive process, then we are dealing with *history*" (46). History, he argues, resides in the body's confrontation with cultural abstraction: it is the process through which the body extends beyond the experience of pleasure, pain, and the realm of sense and translates these experiences into the discursive domains of culture. For Lefèbvre, the history of the body and the history of the senses have been subsumed in modern Western culture into the science and philosophy of social space.

77. All citations from *Antony and Cleopatra* are from Greenblatt et al., *Norton Shakespeare*. As he tries to leave Egypt, Antony foolishly locates Cleopatra's power in her visual beauty: "Would I had never seen her" (1.2.138).

78. On the origins of the Magi tale and its popularity in late medieval Europe, see Marcia Rickard, "The Iconography of the Virgin Portal at Amiens," *Gesta* 22.2 (1983): 149.

79. London's mayoral pageants dispersed spices, sulfurous mists, and aromatic waters into crowds. On the use of scents as stage properties in London's mayoral pageants and on early modern English stages, see Dugan, "Scent of a Woman," 233–34.

80. Jonathan Gil Harris, "'Narcissus in Thy Face': Roman Desire and the Difference It Fakes in *Antony and Cleopatra*," *Shakespeare Quarterly* 45.4 (1994): 408–25, 418 n. 38.

81. Leonard Tennenhouse argues, "Cleopatra is Egpyt," embodying everything that is not Roman (or, for that matter, English); *Power on Display: The Politics of Shakespeare's Genres* (London: Routledge, 1986), 144.

82. See Richmond Tyler Barbour, *Before Orientalism: London's Theatre of the East* (New York: Cambridge University Press, 2003), 58; Janet Adelman, *Suffocating Mothers: Fantasies of Maternal Origin in Shakespeare's Plays: Hamlet to the Tempest* (New York: Routledge, 1992), 178; John Michael Archer, "Antiquity and Degeneration: The Representation of Egypt and Shakespeare's *Antony and Cleopatra*," *Genre: Forms of Discourse and Culture* 27.1–2 (1994): 1–27, 6; and Walter Cohen, "Introduction: *Antony and Cleopatra*," in Greenblatt et al., *Norton Shakespeare*, 2638.

83. Barbara Hodgdon, "Antony and Cleopatra in the Theatre," in *The Cambridge Companion to Shakespearean Tragedy*, ed. Claire McEachern (New York: Cambridge University Press, 2002), 241.

84. Jonathan Dollimore, "Shakespeare, Cultural Materialism, Feminism, and Marxist Humanism," *New Literary History* 21.3 (1990): 471–93, 487.

85. On the use of scents as stage properties, see Dugan, "Scent of a Woman"; see also Harris, "The Smell of Macbeth."

86. At one point, Enobarbus even mistakes Cleopatra *for* Antony (1.2.68–69).

87. Cleopatra's desirability is linked to Egypt's power. Antony invokes the language of enslavement: "These strong Egyptian fetters I must break" (1.2.105).

88. Some critics interpret *Antony and Cleopatra*'s political geography as expressing cultural nostalgia for the Elizabethan reign. See Harris, "Narcissus in Thy Face," 427 n. 38, and Tennenhouse, *Power on Display*, 146.

89. Plutarch's *Lives* has been identified as a key historical source for Shakespeare's *Antony and Cleopatra*. Thomas North's English translation translates this scene as follows:

"[S]he disdained to set forward otherwise, but to take her barge in the riuer Cydnus, the poope whereof was of gold, the sailes of purple, and the owers of siluer, which kept stroke in rowing after the sound of the musicke of flutes, howboyes, citherns, violls, and such other instruments as they played upon the barge. And now for the person of her selfe: she was layed under a pauillion of cloth of gold of tissue, apparelled and attired like the goddesse Venus, commonly drawen in picture: and hard by her on either hand of her, pretie faire boyes apparelled as painters doe set forth god Cupide, with litle fannes in their hands, with which they fanned wind upon her. Her Ladies and gentlewoman also, the fairest of them were apparelled like nymphes Nereides (which are the mermaides of the waters) and like the Graces, some stearing the helme, others tending the tackle and ropes of the barge, out of the which there came a wonderfull passing sweete sauor of perfumes, that perfumed the wharfes side, pestered with innumerable multitudes of people. Some of them followed the barge all alongest the riuers side: others also ranne out of the citie to see her coming in. So that in thend, there ranne such multitudes of people one after another to see her, that *Antonius* was left post alone in the market-place, in his Imperial seate to geue audience"; Plutarch, *The Lives of the Noble Grecians and Romanes Compared Together . . .* (London, 1579), sig. NNNN iiiij v.

Chapter 1 · *Censing God*

1. See Richard Rex, "Which Is Wyche: Lollardy and Sanctity in Lancastrian London," in *Martyrs and Martyrdom in England, c. 1400–1700*, ed. Thomas S. Freeman and Thomas F. Mayer (Rochester, NY: Boydell & Brewer Press, 2007), 88–106, 96.

2. William Gregory, *Chronicle of London*, in *The Historical Collections of a Citizen of London in the Fifteenth Century*, ed. James Gairdner ([Westminster, U.K.]: Camden Society, 1876), 183, cited in Rita Copeland, *Pedagogy, Intellectuals, and Dissent in the Later Middle Ages: Lollardy and Ideas of Learning* (New York: Cambridge University Press, 2001), 188.

3. Paul Murray Kendall, *The Yorkist Age: Daily Life during the War of the Roses* (New York: W. W. Norton, 1962), 276.

4. Ibid.

5. A. H. Thomas and I. D. Thornely, eds., *The Great Chronicles of London* (London: G. W. Jones, 1938), 175.

6. See Rex, "Which Is Wyche," 93.

7. See Thomas and Thornley, *The Great Chronicle*, 175.

8. For more on Virley, see Rex, "Which Is Wyche," 101.

9. I pun on Richard Rex's clever title, "Which Is Wyche?"

10. William Caxton, *The Cronycles of Englond* (London, 1482), sig. X2 v.

11. Historian Susan Harvey argues that early Christians "used olfactory experience to formulate religious knowledge" and demonstrates how incense, an earthly scent associated with embalming and sacrifice, came to represent the smell of the divine. Until the fourth century CE, between 1,300 and 1,700 tons of frankincense (approximately 7,000–10,000 camel loads) were imported annually into Rome; *Scenting Salvation: Ancient Christianity and the Olfactory Imagination* (Berkeley & Los Angeles: University of California Press, 2006), 35. Yet incense's symbolic importance increased after the fourth century when its scent betrayed a rarified, experiential understanding of divinity (rather than a common scent of sacrifice), one linked to church spaces and scriptural dogma. Its powerful scent offered a direct, experiential encounter with god, amid the increasingly multisensorial liturgy.

12. Incense was a trope that united the Passion and Easter liturgical celebrations. Passion plays usually dramatized the conversion of Mary Magdalene, the Last Supper, and ultimately, the Passion of Christ. The conversion scenes staged Mary Magdalene, under the influence of Diabolus, procuring scented balms and unguents to anoint her body and entice lovers. The Easter liturgy also emphasized the Magdalene's visit to the apothecary, elaborating on these aspects of the tale. See Karl Young, *The Drama of the Medieval Church*, 3 vols. (Oxford: Claredon, 1933), 1:236.

13. See Roger E. Reynolds, "The Drama of Medieval Liturgical Processions," *Revue de Musicologie* (2000): 127–42, 137–38.

14. In his analysis of Tudor and Stuart England, David Cressy argues that incense also fulfilled a practical function: combating the smell of human decomposition associated with internment rituals; *Birth, Marriage, and Death: Ritual, Religion, and the Life-Cycle in Tudor and Stuart England* (New York: Oxford University Press, 1997), 463.

15. The chronicler Richer of Saint Denis tells of a woman who claimed prophetic powers, demonstrated by her ability to subsist on angelic spices while locked in a cell. It was later revealed that food was smuggled to her and that spices were scattered to mimic the odor of sanctity. See Paul Freedman, *Out of the East: Spices and the Medieval Imagination* (New Haven, CT: Yale University Press, 2008), 84.

16. See Michael Staunton, *Thomas Becket and His Biographers* (Woodbridge, U.K.: Boydell Press, 2006), 211.

17. See C. M. Woolgar, *The Senses in Late Medieval England* (New Haven, CT: Yale University Press, 2006), 118.

18. *Etymologies,* I: lib. 5.31.1, cited in Woolgar, *Senses,* 119.

19. Roger Dymok, "Against the XII Heresies of the Lollards," in *English Historical Review* 22 (1907): 295–304, 300.

20. See Ronald C. Finucane, *Miracles and Pilgrims: Popular Beliefs in Medieval England* (Basingstoke, U.K.: Palgrave Macmillian, 1995), 200.

21. Misattributed to Henry Parker, *Dives and Pauper* (London, 1536), commandment 1, chap. 15. For an analysis of this text, see Michael Camille, "The Iconoclast's Desire: Deguileville's Idolatry in France and England," in *Images, Idolatry, and Iconoclasm in Late Medieval England: Textuality and the Visual Image,* ed. Jeremy Dimmick (New York: Oxford University Press, 2002).

22. Erasmus, *An Exposition of the XV Psalm* (London, 1537), n.p.

23. Thomas Becon, *A Fruitful Treatise of Fasting* (London, 1551), sig. F i r–v. For the relationship between censing and funerals, see Cressy, *Birth, Marriage, and Death,* 463.

24. *The Riverside Chaucer,* 3rd ed., ed. Larry Dean Benson (Boston: Houghton Mifflin, 1987), 1:3339–43. Absolon's liturgical censing is contrasted earlier by more overtly erotic scents–Nicholas's aromatic bedroom perfumes. Nicholas's chamber is "ful festily dyight with herbes swoote, / And he himself as sweete as is the roote / Of licoris or any cetewale"; 1:3205–7.

25. Luther quotes Mantuan in his late work, *Against the Roman Papacy, An Institution of the Devil.* See Lee Piepho, "Mantuan's Eclogues in the English Reformation," *Sixteenth Century Journal* 25.3 (1994): 623–32, 624–25.

26. See E. G. Cuthbert F. Atchley, *A History of the Use of Incense in Divine Worship* (London: Longmans, Green, 1909), 329–30.

27. Jonathan Gil Harris, "The Smell of Macbeth," *Shakespeare Quarterly* 58.4 (2007): 465–86, 484.

28. Ibid., 485.

29. Carol Litzenberger, *The English Reformation and the Laity* (New York: Cambridge University Press, 2002), 116.

30. Ibid.

31. James Peller Malcolm, *Londinium Redivivum: or An Antient History and Modern Description of London* (London: J. Nichols and Son, 1805), 2:88, and *Churchwardens' Accounts of Great Wigston, Leicestershire,* both cited in Thomas North, *A Chronicle of the Church of S. Martin in Leicester* (London: Bell & Daldy, 1866), 127.

32. William Prynne, *Canterburies Doome* (London, 1646), 74, 123, cited in North, *Chronicle,* 127.

33. See North, *Chronicle,* 127.

34. Luther quotes from Mantuan's *De Calamitatibus Temporum.* See Piepho, "Mantuan's Eclogues," 625.

35. See Bale, *Illustrium maioris Britanniae Scriptorium* (1548), f. 154 v, cited in Margaret Aston, *Lollards and Reformers: Images and Literacy in Late Medieval Religion* (New York: Continuum, 1984), 245.

36. "And the sweete smoke of the odorous incense, which came of ye wholsome and feruent desires of them that had fayth, ascended vp before God out of ye Angels hande. By his onely meryte was their fayth accepted, and for his deaths sake their works pleased God. The sayd Angel tooke the censer, he prepared his godlie spirite. He filled it with fire of the aulter, whych was his eternall charitie. And he cast it into the earth, with poure he sent it downe in clouen fyerie tounges vpon hys Apostles, of whose plentuous aboundaunce all wee haue receyued"; John Bale, *The Image of Both Churches after the Most Wonderful and Heauenly Reuleation of Sainct John the Euangelist . . .* (London, 1570), sig. Oiii r–v.

37. John Bale, *The First Two Partes of the Actes or Unchast Examples of the Englysh Votaryes* (London, 1551), sig. Ai v.

38. George Herbert, *A Priest to the Temple, or, The Country Parson* (London, 1652), sig. C3 v–sig. C4 r.

39. Ibid., D5r.

40. See Theophrastus, *Enquiry into Plants and Minor Works on Odours and Weather Signs*, ed. and trans. Arthur Holt, 2 vols. (New York: G. P. Putnam's Sons, 1916), 2:327, cited in Harvey, *Scenting Salvation*, 31.

41. Harvey, *Scenting Salvation*, 14.

42. Ibid.

43. M. Regert, T. Devièse, A. S. Le Hô, and A. Rougeulle, "Reconstructing Ancient Yemeni Commercial Routes during the Middle Ages Using Structural Characterization of Terpenoid Resins," *Archaeometry* 50.4 (2008): 668–95, 670.

44. See Arthur O. Tucker, "Frankincense and Myrrh," *Economic Botany* 40.4 (1986): 425–33, 425.

45. See the University of California at Los Angeles online biomedical library on botany: http://unitproj.library.ucla.edu/biomed/spice/index.cfm (visited September 16, 2004).

46. In his analysis of biblical epiphanies, Rainer Warning argues that there is a profound sensory switch between the Old and New Testament. In Warning's reading, the epiphanies of the Old Testament emphasize visual transcendence, accompanied by visions of burning bushes, storms, and columns of fire, whereas New Testament epiphanies were accompanied by sonic sensory shifts. See "Seeing and Hearing in Ancient Medieval Epiphany," in *Rethinking the Medieval Senses: Heritage, Fascinations, and Frames*, ed. Stephen G. Nichols, Andreas Kablitz, and Alisoun Calhoun (Baltimore, MD: Johns Hopkins University Press, 2008), 102–16. Hans Ulrich Gumbrecht, however, warns against putting too much stock in such large-scale historical arguments; "Erudite Fascinations and Cultural Energies: How Much Can We Know about the Medieval Senses," in Nichols, Kablitz, and Calhoun, *Rethinking the Medieval Senses*, 1–10, 10.

47. In York, the goldsmiths staged the cycle's offering play, which undoubtedly emphasized visual splendor of the epiphany and the gift of gold; however, the use of this motif would also require attention to the offerings of myrrh and incense. In Chester, the mercers and the vintners staged the two plays of the Magi, again highlighting the wealth of the guilds with the wealth of the kings' offerings. For more on this link, see Christina M. Fitzgerald, *The Drama of Masculinity and Medieval Guild Culture* (Houndsmill, U.K.: Palgrave Macmillan, 2007), 112.

48. For more on the interpretive potential latent in such heterogeneity, see Ananya Jahanara Kabir and Deanne Williams's introduction in *Postcolonial Approaches to the European Middle Ages* (New York: Cambridge University Press, 2005), 1–24, 5.

49. Robert of Torigni, an English witness, described their holy and earthly bodies: "Their outer bodies remained intact so far as the skin and hair because they had been buried with balsam and other unguents. The first of the Magi . . . appeared to be fifteen years old, the second thirty, and third, sixty"; Joseph P. Huffman, *The Social Politics of Medieval Diplomacy: Anglo-German Relations* (Ann Arbor: University of Michigan Press, 2000), 73. The miracle of their preservation stemmed from earthly funeral practices, linked to the scented balms they offered to the infant Christ, even as the description makes the Magi into allegorical representations of humanity.

50. "The magi have power and personality only as a collectivity and almost never as individuals. They were not recognized as saints. There was no significant cult to Balthasar,

to Melchior, or to Caspar. . . . [I]t is fair to say that in general, the faithful at the end of the eleventh century were more likely to pray like the magi than to pray to them"; Richard C. Trexler, *The Journey of the Magi* (Princeton, NJ: Princeton University Press, 1997), 73.

51. Patrick J. Geary, *Living with the Dead in the Middle Ages* (Ithaca, NY: Cornell University Press, 1994), 247.

52. Ibid., 251.

53. Mathew of Paris claimed that when the Mongols attacked Christendom in 1243, their motivation was to recover the bodies of the Magi. See Crombach, *Primitiae Gentium, sen Historia* . . . (Cologne, 1654), 164, cited in Texler, *Journey of the Magi*, 74.

54. Bede, *In Mattaei Evangelium Exposito*, in *Patrologia Latina*, ed J. Migne, 9–132, 13–15, cited in Paul H. D. Kaplan, *The Rise of the Black Magus in Western Art* (Ann Arbor, MI: UMI Research Press, 1985), 26.

55. Jacobus de Voragine, *The Golden Legend: Readings on the Saints*, vol. 1, trans. William Granger Ryan (Princeton, NJ: Princeton University Press, 1995), 82–3.

56. For more, see Pamela M. King, *York Mystery Cycle and Worship of the City* (Cambridge, U.K.: D. S. Brewer, 2006), 110.

57. All references from the Wakefield *Offering of the Magi* are from *Medieval Drama*, ed. David Bevington (Boston: Houghton Mifflin, 1975), 409–28.

58. All references from the *Second Shepherds' Play* are from *Early English Drama*, ed. John Coldewey (New York: Garland, 1993), 343–63.

59. Gib draws on two proverbs to explain the animal like features that are "marked amiss" on the couple's "barne"; both pun on Gyll's sexuality. The first–"ill-spun weft, iwys, ay commys foull owte" (line 587)–emphasizes the weaving work she performs in the home, linking home economies with sexual ones. The second–"I trowe kynde will crepe / where it may go" (line 591)–suggests that the child's sheeplike features are the result of Gyll's bestial couplings.

60. Stephen Mead, "Four-Fold Allegory in the Digby *Mary Magdalene*," *Renascence* 43.4 (1991): 269–82, 279.

61. Though loose hair often signifies virginal innocence in medieval visual traditions, the Magdalene's loose hair is linked to her conversion and thus her previous sexual sins. See Katherine Janse, *The Making of the Magdalene* (Princeton, NJ: Princeton University Press, 2000), 134.

62. Scholars debate whether the Digby *Mary Magdalene* reflects pre- or post-Reformation attitudes toward the saint. The manuscript was transcribed by Myles Bloomfield in the first quarter of the sixteenth century, but its language suggests that it was probably first performed in the late fifteenth century. Although there is a lack of evidence about the play's performance history, it could have been written and performed after 1518, when Jacques Lefèvre d'Étaples published his attack on the Magdalene's composite biography and St. John Fisher responded with a defense. See Clifford Davidson, "The Digby *Mary Magdalene* and the Magdalene Cult of the Middle Ages," *Annuale Mediavele* 13 (1972): 70–87, 71. Most scholars locate the writing of the play somewhere between 1480 and 1530. Theresa Colleti argues that the play is East Anglian; *Mary Magdalene and the Drama of Saints: Theater, Gender, and Religion in Late Medieval England* (Philadelphia: University of Pennsylvania Press, 2004), 36.

63. Theresa Coletti, "*Paupertas Est Donum Dei*: Hagiography, Lay Religion, and the Economics of Salvation in the Digby *Mary Magdalene*," *Speculum: A Journal of Medieval Studies* 76.2 (2001): 337–78, 313. In a subsequent study of this play, Coletti argues that it "exhibits thematic preoccupations and theatrical images that, viewed collectively, stunningly resonate with the spiritual ideology and activities of the late medieval hospital"; *Mary Magdalene and the Drama of Saints*, 39.

64. Her feast day, July 22, marks the anniversary of her death, when she was swept into a church nearby, received Holy Communion, and died at the altar. According to Gregory's account, the church smelled of perfume for seven days, signaling her arrival in heaven. See Susan Haskins, *Mary Magdalene: Myth and Metaphor* (New York: Riverhead Books, 1993), 104, and Katherine Ludwig Jansen, *Making of the Magdalene: Preaching and Popular Devotion in the Later Middle Ages* (Princeton, NJ: Princeton University Press, 2001), 35–36.

65. Glynne Wickham, *Early English Stages, 1300–1600* (New York: Routledge, 1959–72), 1:90.

66. Mead, "Four-Fold Allegory," 270.

67. See Coletti, *Mary Magdalene and the Drama of Saints*, 156–58.

68. Theresa Coletti, "'Curtesy Doth It Yow Lere': The Sociology of Transgression in the Digby *Mary Magdalene*," *English Literary History* 71.1 (2004): 1–28, 2.

69. Donald C. Baker, John L. Murphy, and Louis B. Hall Jr., eds., *The Late Medieval Religious Plays of the Bodleian Mss. Digby 133 and E Museuo 160* (New York: Oxford University Press, 1982), 35, lines 335–45.

70. Lady Lechery replies, "O ye prynse, how I am ful of ardent lowe, / Wyth sparkyllys ful of amerowsnesse! / Wyth yow to rest fayn wold I aprowe, / To shew plesavns to your jentylnesse!"; ibid., lines 352–55.

71. Galingale resembles a mild gingerroot both in taste and smell and was very popular in medieval cookery. It is found both in southern Egypt and southeast Asia. Grains of paradise, an African spice, resemble a cross between pepper, ginger, and cardamom; clary is a type of sage found in England; marjoram is a sweet spice native to northern Europe.

72. See Deanne Williams, *The French Fetish from Chaucer to Shakespeare* (New York: Cambridge University Press, 2004), 71. For an analysis of sodomy in the late medieval period, see Bruce Holsinger, "Sodomy and Resurrection: The Homoerotic Subject of the *Divine Comedy*," in *Premodern Sexualities*, ed. Louise Fradenburg and Carla Freccero (New York: Routledge, 1996): 243–74, and Carolyn Dinshaw, *Getting Medieval: Sexualities and Communities, Pre- and Postmodern* (Durham, NC: Duke University Pres, 1999). See also Mark Jordan, *The Invention of Sodomy in Christian Theology* (Chicago: University of Chicago Press, 1997).

73. Theresa Coletti notes that these "queer moments of dramatic action have gone largely unremarked in scholarship." Coletti focuses on masculinity "as a preeminent dramatic locus of erotic energy and behavior," analyzing the pagan priest of Marseille's relationship with his clerk; the shipmaster, Nauta, and his boy; and, finally, the King of Marseille and Nauta; *Mary Magdalene and the Drama of Saints*, 159–61.

74. Baker, Murphy, and Hall, *Late Medieval Religious Plays*, 26, lines 71–72.

75. For a discussion of contrapasso, see Jansen, *Making of the Magdalene*, 156.

76. Christ tells the Magdalene, "Wan that gardyn [of virture] is watteryed wyth teyrs clere, Than spryng vertuus, and smelle full sote"; Baker, Murphy, and Hall, *Late Medieval Religious Plays*, lines 1356–65.

77. The first three amplify incense's soft scent, while the last stages her divinity through a distinctively English scent.

78. Baker, Murphy, and Hall, *Late Medieval Religious Plays*, lines 1356–65.

Chapter 2 · Casting Selves

1. Edward Hall, *The Union of Two Noble and Illustre Famelies of Lancastre and Yorke* (London, 1548), fol. lxxxvi, sig. PPpvi, r–v.

2. Ibid, fol. xxxxif, sig. QQqii r.

3. Ibid.

4. Ibid.

5. Episode 4, of the first season, "His Majesty, the King" stages this masque. It first aired on Showtime April 22, 2007.

6. "Inde" connoted both the location of India and the products associated with it.

7. Glynne Wickham, *Early English Stages, 1300–1600*, 3 vols. (New York: Routledge, 1959–72), 1:96.

8. Francis Bacon, "Of Gardens," in *The Works of Francis Bacon*, ed. James Spedding and Robert Ellis (London: Longman, 1878), 487.

9. Erasmus, *Collected Works of Erasmus: Colloquies*, trans. Craig R. Thomson (Toronto: University of Toronto Press, 1997), 265.

10. See Howard M. Colvin, "Royal Gardens in Medieval England," in *Medieval Gardens*, ed. Elisabeth B. MacDugall (Washington, DC: Dumbarton Oaks, 1986), 7–23, 12–13.

11. Rose Standish Nichols, *English Pleasure Gardens* (New York: Macmillan, 1902), 107.

12. David Hume and Tobias George Smollett, *Hume and Smollett's Celebrated History of England*, ed. John Robinson (Hartford: D. F. Robinson, 1827), 493.

13. Jacob Burckhardt, *Civilization of the Renaissance in Italy*, ed. Peter Murray, trans. S.G.C. Middlemore (New York: Penguin Classics, 1990), 127.

14. Rediscovered in the thirteenth century, it is believed that French knight Robert de Brie, returning home from the seventh crusade, brought a specimen to Champagne. Though the distinction is often given to *rosa centifolia,* it was undoubtedly hundreds of camel-loads of rosewater distilled from damask roses that Saladin used to purge the mosque of Omar of Christianity when he reclaimed Jerusalem in 1193 CE. Rimmel retells this legend in his description of *rosa centifolia;* however, the cabbage rose is native to southern Europe, though many believe it to be a hybrid, first grafted in the early sixteenth century by Dutch gardeners for perfume. See Frances Parkman, *The Book of Roses* (Boston: J.E. Tilton, 1866), 120.

15. See Ralph Austen, *Observations upon Some Part of Sr Francis Bacon's Naturall History* (Oxford, 1658), sig. F3 r.

16. All references to Shakespeare's sonnets are from *The Norton Shakespeare: Based on the Oxford Edition*, 2nd ed., ed. Stephen Greenblatt, Walter Cohen, Jean E. Howard, and Katharine Eisaman Maus (New York: W. W. Norton, 2008), 1946–99.

17. Her legacy lives on in the alchemical tools it is believed that she designed, namely the bain-marie. For more on Maria the Jewess, see Naomi Janowitz, *Magic in the Roman World: Pagans, Jews, and Christians* (New York: Routledge, 2001), 61.

18. None of the writings of Maria the Jewess survive, only quotes in the writings of Zosimus of Panapolis. See Stephen Edelston Toulmin and June Goodfield, *The Architecture of Matter* (Chicago: University of Chicago Press, 1982), 122. These practices came to be described in European literature as "alchemy" in the twelfth century. By the thirteenth century, however, distillation was associated with the Franciscan friars, due to the work of John of Rupescissa, a Spanish monk, who first connected alcohol with Aristotelian science, and Raymond Llull, a Catalonia missionary.

19. See Barbara Sebek, "Canary, Bristoles, Londres, Ingleses: English Traders in the Canaries in the Sixteenth and Seventeenth Centuries," *A Companion to the Global Renaissance: English Literature and Culture in the Era of Expansion*, ed. Joystna Singh (Malden, MA: Wiley-Blackwell, 2009), 279–93.

20. Leo Africanus, *A Geographical Historie of Africa*, trans. John Pory (London, 1600), sig. N6 r.

21. Two other texts are mentioned: a "huge volume of the same arte, intituled by the name of Attogrehi," along with "the songs or articles of the said science were written by one Mugairibi of Granada, whereupon a most learned Mamaluch of Damasco wrote a commentarie; yet so, that a man may much more easily vnderstand the text then the exposition thereof" (ibid.).

22. Sebastian Munster and Rychard Eden, *A Treatyse of the New India* (London, 1553), sig. Diiij v.

23. Thomas Nashe, *The Unfortunate Traveller. Or, the Life of Jacke Wilson* (London, 1594), sig. F r.

24. Hugh Plat, *Delight for Ladies* (London, 1609), sig. G9 v.

25. Bruce T. Moran, *Distilling Knowledge: Alchemy, Chemistry, and the Scientific Revolution* (Cambridge, MA: Harvard University Press, 2005), 25.

26. David Michael Stoddart, *The Scented Ape: The Biology and Culture of Human Odor* (New York: Cambridge University Press, 1990), 149.

27. See Edward Sagarin, *The Science and Art of Perfumery* (New York: McGraw-Hill, 1945), 120.

28. Sue Clarke, *Essential Chemistry for Aromatherapy*, 2nd ed. (Philadelphia: Churchill, Livingston, Elsevier, 2009), 187–88.

29. Mandy Aftel, *Essence and Alchemy* (Layton, UT: Gibbs Smith, 2001), 109.

30. See "The Lord Bacon's Medical Remedies," in *Baconia* (London, 1679), 158.

31. See Heinrich von Staden, *Herophilus: The Art of Medicine in Early Alexandria* (New York: Cambridge University Press, 1989), 540–54, and Lise Manniche, *Sacred Luxuries: Fragrance, Aromatherapy and Cosmetics in Ancient Egypt* (Ithaca: Cornell University Press, 1999), 80.

32. Paulus's texts were translated into French and Latin many times in the sixteenth century. They were not translated into English until the middle of the nineteenth century. He writes, "[T]here is one class of medicines possessed of a similar temperament to our bodies . . . and another that is of a hotter temperature than we. Of this temperament I have thought it right to make four orders, the first being imperceptible to the senses, and only to be inferred from reflection; the second being perceptible to the senses; the thing strongly heating but not burning; and the fourth, or last, caustic"; *The Seven Books of Paulus Aeginita,* trans. Francis Adams (London: Sydenham Society, 1847), vol. 3, chap. 7, 6.

33. Ibid.

34. See "rosewater" in the "Table of Simples" in John Banister's *A Needefull, New, and Necessarie Treatise of Chyrurgerie* (London, 1575), n.p.

35. In Greek mythology, roses were created by Chloris, who stumbled upon a beautiful, lifeless nymph in a forest. Chloris transformed the nymph into a flower, Aphrodite gave it a gift of beauty, and Dionysus provided its aromatic nectar. Leonhart Fuchs, *The Great Herbal of Leonhart Fuchs: De Historia Stirpium Commentarii Insignes, 1542,* ed. Frederick G. Meyer, Emily Emmart Trueblood, and John L. Heller (Palo Alto, CA: Stanford University Press, 1999), 612.

36. See Samuel Bowne Parsons, *Parsons on the Rose: A Treatise on the Propagation, Culture, and History of the Rose* (New York: Orange Judd, 1869), 187.

37. R. A. Donkin, *Dragon's Brain Perfume: A Historical Geography of Camphor,* Brill's Indological Library, vol. 14 (Leiden: Brill, 1999), 17.

38. Susan Harvey, *Scenting Salvation: Ancient Christianity and the Olfactory Imagination* (Berkeley & Los Angeles: University of California Press, 2006), 118.

39. In comparison, John Okborn, a clerk and vicar of St. Dunstan-next-Canterbury, left ten shillings in his will to repair the church of St. Dunstan; Consistory Court of Canterbury, vol 2., folio 97, cited at www.ogbourne.com/wjok1456.htm. See Cov. Leet book 292, cited in *Oxford English Dictionary,* s.v. "rosewater."

40. Cited in Bruce T. Moran, *Distilling Knowledge,* 63.

41. Remnants of distilling, including tubing and glass shards of receivers and cucurbits, the lower part of an alembic, have been found at a number of sites, suggesting that most monasteries had full-blown "still" rooms. See W. John Blair and Nigel Ramsay, *English Medieval Industries* (New York: Continuum International, 2001), 60.

42. "The lower Region of the Air is like unto the neck or higher part of an Alembic, for through it the Vapours climbing up, and being brought to the top, receive their condensation from Cold. . . . In these operation of Nature, the Earth is the Vessell receiuing"; Jean d'Espagnet, *Enchyridion Physicae Restitutae: or the Summary of Physics Recovered* (London, 1651), p. 19, cited in Allen George Debus, *The Chemical Philosophy: Paracelsian Science and Medicine in the Sixteenth Century* (New York: Dover, 2002), 88. See also Debus, *The French Paracelsians: The Chemical Challenge to Medical and Scientific Tradition in Early Modern France* (New York: Cambridge University Press, 2002), 25, 63.

43. Conrad Gesner's *Thesaurus Euonymi (Part i)* was translated in 1559; the second part was translated in 1576. Even Paré added an appendix on distillation in the 1580s. For more, see Debus, *French Paracelsians,* 13.

44. Hieronymus Brunschwig, *The Vertuose Boke of Distyllacyon . . .* , trans. Laurens Andrewe (London, 1527), sig. Aii r.

"Dystyllynge is necessaryly founde and ordeined for many maner of necessytees / and specyally for the loue of man hym for to kepen helthe & strength and to brynge the seke and weke body agayne to helthe / and to the entent that the grosse and corruptyble body may be agayne clensed and puryfyed / for who soeuer taketh herbes / roses / or other substances and stampe them / the juice thee of strayned and mynstred is not lightely whiche many one therfore doth them abhorre bycawse of the inconnenyent syght. Secondaryly with waters dystyllyd all maner of confeccyons / syropys / powders & electuaryes be mixed to the entet that they should be the more syghtly and doulcet to be minstred receyued and used also this dystylling is onely founde for the comon people that dwelle farre frome medycynes & physycyons & for them that ben not able to paye for costely imedycnes / the whiche hath moued me greatly this my lyteil worke to opene and dysclose for the helthe and prosperyte of myne euen crysten. Thyrdely. the dystyllynge is ordeyned bycause that whan one medycne is mynystred with the corpus or substaunce in the maner of electuaryes / confeccyons powders or syropys or any medycynes with etynge or swalowynge downe / or drynkyng in what maner so euer it be is people" (sig. Ai r).

45. Ibid, sig. Riii r–Si v.

46. Ibid., sig. Riv r.

47. The *Treasure of Euonymus* uses rosewater to describe what "distillation is" as well as the "vertues" of "licores distilled generally." *Newe Jewell of Health* uses it to describe the processes of making both vinegar and hemp water and provides numerous recipes that use it as a central ingredient. Konrad Gesner, *Treasure of Euonymus* (London, 1559), sig. B r, Biv r; *Newe Jewell of Health* (London, 1576), sig. Ci r, Ciiij r.

48. See Antonio Guevara, *A Booke of the Inuention of the Art of Nauigation* (London, 1578), sig. Oy r.

49. Parsons, *Parsons on the Rose,* 204.

50. Michael Clayton, *Collectors Dictionary of the Silver and Gold of Great Britain and North America* (New York: World, 1971), 50.

51. See, for example, Patricia Berrahou Phillippy's reading of this casting bottle in *Painting Women: Cosmetics, Canvases, and Early Modern Culture* (Baltimore, MD: Johns Hopkins University Press, 2006), 134–35.

52. William Bullein, *Bulwarke of Defence Against All Sicknesse, Soarenesse, and Woundes That Do Dayly Assaulte Mankind* (London, 1572), fol. 57, sig. Riii r.

53. Ibid.

54. Geoge Gascoigne, *A Hundreth Sundrie Flowers Bounde Up in One Small Poesie* (London, 1573), sig. Nii r–v, 261–62.

55. Francesco Colonna, *Hypnerotomachia,* trans. Robert Dallington (London 1592), sig. L3 r–v, 39.

56. John Florio, *A Worlde of Wordes, or Most Copious, and Exact Dictionarie in Italian and English* (London, 1598), 4, 249.

57. Nathan Field, *A Woman's a Weather-cocke* (London, 1612), sig. E2 r. See also the entry for "bottle" in Gordon Williams, *A Dictionary of Sexual Language and Imagery in Shakespearean and Stuart Literature* (London: Athlone, 1994), 1:135.

58. Henry Chettle, *The Tragedy of Hoffman, or a Revenge for a Father* (London, 1631), Actus secundus, sig. C4 r.

59. *Hamlet,* in Greenblatt et al., *Norton Shakespeare,* 1683–1784.

60. See Michael Schoenfeldt's discussion of George Herbert in *Bodies and Selves in Early Modern England: Physiology and Inwardness in Shakespeare, Spenser, and Herbert* (Philadelphia: University of Pennsylvania Press, 1999), 126.

61. Richard Halpern, *Shakespeare's Perfume: Sodomy and Sublimity in the Sonnets, Wilde, Freud, and Lacan* (Philadelphia: University of Pennsylvania Press, 1999), 14.

62. "And to his robbery had annex'd thy breath" (lines 8–11).

63. "Venus and Adonis," in Greenblatt et al., *Norton Shakespeare,* 629–68.

64. Herbert Appold Greuber, *Handbook of the Coins of Great Britain and Ireland in the British Museum* (London: British Museum, 1899), 75.

65. See John Gerard, *Herball, or Genreal Historie of Plants* (London, 1633), sig. Aaaaaa v, 1387.

66. *Sir Philip Sydney,* ed.Katherine Duncan-Jones, Oxford Authors series (New York: Oxford University Press, 1989), 303–4, cited in Robert Matz, *Defending Literature in Early Modern England: Renaissance Literary Theory in Social Context* (New York: Cambridge University Press, 2000), 79.

67. The significance of this association is made clear in a less famous exchange between the queen, an accused Catholic priest, and the queen's interrogator, Richard Topcliffe, who purportedly was overly familiar with the queen's body, declaring it "the softest belly of any womankind." The queen replied, "be not these the arms, legs, and body of King Henry?," to which he answered, "yea." See Christopher Pye, *The Regal Phantasm: Shakespeare and the Politics of Spectacle* (New York: Routledge, 1990), 34.

68. Steven Mullaney, *The Place of the Stage* (Chicago: University of Chicago Press, 1988).

69. There is a dangerous supercessionary logic to this transformation; London's memorial to the Jewish cemetery is "presumably" unused, "one must imagine, since Edward's expulsion of the Jews in England," and thus reconfigured into a theater space. See Mullaney, *Place of the Stage,* 1. For a theory of supercessionary logic, see K. Biddick, *The Typological Imaginary: Circumcision, Technology, History* (Philadelphia: University of Pennsylvania Press, 2003), and Jonathan Gil Harris, *Untimely Matter in the Time of Shakespeare* (Philadelphia: University of Pennsylvania Press, 2008), esp. chap. 1 and 2.

70. See Ian Munro, *The Figure of the Crowd: Early Modern London and Its Double* (Houndsmill, U.K.: Palgrave Macmillan, 2005), 4.

71. Roses had long signified brothels in European cities; medieval Catholic iconography of the Virgin symbolized by a pure, white rose lent a useful pictorial language to red ones, which were a symbol of women's plucked "maidenhead," referring to profligate women and covertly advertising prostitutes. London, like many European cities, had a number of brothels with roses in their name. Such associations worked in visual and tactile ways: the color of the red rose served as a shorthand for the "plucking" that took

place inside. Yet the presence of the two rose gardens in the 1540s suggests that there also was an olfactory component to such associations, and that the meanings of such associations were complicated.

72. *Stews,* after all, were stocked ponds used by fishmongers, which explains how one Bankside industry extended to another. See Richard Rex, *"The Sins of Madame Eglantine" and Other Essays on Chaucer* (Dover: University of Delaware Press, 1995), 88. Like the nearby ponds stocked with fish, the brothels of Bankside were stocked to meet consumer appetites, albeit differently. Stews also referred to stoves, cauldrons, or the spaces that contained them, linking topographical associations with metonymic ones, especially when applied to public space. It did not take much for stews, or heated public baths, to become associated with vices that occurred there. Finally, stews connoted a suffocating vapor or stench that redefines the space in a new way. Thus the term *stew* was a mobile placeholder for a host of sensory associations.

73. See Ernest L. Rhodes, *Henslowe's Rose: The Stage and Staging* (Louisville: University of Kentucky Press, 1976), 1–2.

74. See E. K. Chambers, *The Elizabethan Stage* (New York: Oxford University Press, 2009), 2:405. Though *garden* was a term that could also refer to fishponds, the name of "Rose alley" emphasizes the area's horticultural gardens. For more, see Julian Bowsher and Pat Miller, *The Rose and the Globe: Playhouses of Shakespeare's Bankside, Southwark, Excavations 1988–90* (London: Museum of London Archeology, 2009), 14–16.

75. See Stephen Greenblatt, *Will in the World: How Shakespeare Became Shakespeare* (New York: W. W. Norton, 2004), 181.

76. See Richard West, *The Court of Conscience* (London, 1607), quoted in Thornton S. Graves, "Some References to Elizabethan Theaters," *Studies in Philology* 19.3 (1922): 317–27.

77. 3.1.4–4, quoted in Jason Scott-Warren, "When Theaters Were Bear-Gardens; Or, What's at Stake in the Comedy of Humors," *Shakespeare Quarterly* 54.2 (2003): 63–83, 67.

78. In the induction, the author describes the stage as "dirty as Smithfield and stinking every whit"; Ben Jonson, *Bartholomew Fair,* in *Renaissance Drama,* ed. Arthur Kinney (London: Blackwell, 2000), 481–555.

79. All citations are from Thomas Dekker, "Shoemaker's Holiday," in Kinney, *Renaissance Drama,* 243–86.

80. See Kathleen E. McLuskie, "Shakespeare's 'Earth-Treading Stars': The Image of the Masque in Rome and Juliet," *Shakespeare Survey* 24 (1971): 63–69, 63.

81. Catherine Belsey, "The Name of the Rose in *Romeo and Juliet,*" *Yearbook in English Studies* 23 (1993): 126–42, 126.

82. Ibid., 127.

83. All quotations from *Romeo and Juliet* are from Greenblatt et al., *Norton Shakespeare,* 905–72.

84. Belsey, "Name of the Rose," 137.

85. See, for example, Tanya Pollard's reading of this scene in *Drugs and Theatre in Early Modern England* (New York: Cambridge, 2005), 58.

Chapter 3 · Discovering Sassafras

1. See "A Discourse Written by Sir Humphrey Gilbert, Knight" in Henry Morley, ed., *Voyages in Search of the North-west Passage: From the Collection of Richard Hakluyt* (London: Cassell & Company, 1886), 41.

2. Gilbert's motto takes on a sinister air when considered alongside his actions as military commander in Ulster during the Irish wars. Gilbert reported in a letter to Sir Henry Sydney that "I slew all those from time to time that did belong to, feed, accompany, or maintain any outlaws or traitors; and after my first summoning of any castle or fort, if they would not presently yield it I would not afterwards take it of their gift, but won it perforce, how many lives so ever it cost, putting man, woman and child of them to the sword"; quoted in Ben Kiernan, *Blood and Soil: A World History of Genocide from Sparta to Darfur* (New Haven, CT: Yale University Press, 2007), 47. As David Quinn concluded, Gilbert's "method of waging war was to devastate the country, killing every living creature encountered"; *Voyages and Colonising Enterprises of Sir Humphrey Gilbert* (London: Hakluyt Society, 1940), 1:17, also cited in Kiernan, *Blood and Soil*, ibid.

3. George Gascoigne, "The Epistle to the Reader," in Sir Humfrey Gilbert, *A Discourse of a Discouerie for a New Passage to Cataia* (London, 1576), sig. qiiij r.

4. Ibid., sig. qqiiij v.

5. In 1497, Henry VII granted a patent to explore westward lands, including Ireland. John Cabot, hoping to discover China, discovered Newfoundland and its early fishing colonies.

6. Gilbert got as far as Ireland, engaging in acts of piracy against Spain. This was Gilbert's second endeavor to the new world; the first ended disastrously. See Kim Sloan, "Setting the Stage for John White, A Gentleman in Virginia," in *A New World: England's First View of America,* ed. Kim Sloan, (Chapel Hill: University of North Carolina Press, 2007), 17, 19. The reward for such expertise included large tracts of land in the St. Lawrence area, including one granted to John Dee. See Gillian Cell, *English Enterprise in Newfoundland* (Toronto: University of Toronto Press, 1969), 41.

7. See Sloan, "Setting the Stage," 21.

8. When Gilbert arrived in St. John's Bay, Newfoundland, sixteen Spanish, Portuguese, and French fishing ships were already in the harbor. Nevertheless, on August 5, 1583, Gilbert formally took possession of Newfoundland for the English. See Cell, *English Enterprise,* 23, and Sloan, "Setting the Stage," 21. For more on the early European fishing industry in Newfoundland, see Peter Edward Pope, *Fish into Wine: The Newfoundland Plantation in the Seventeenth Century* (Chapel Hill: University of North Carolina Press for the Omohundro Institute of Early American History & Culture, 2004).

9. Based on reports by Frobisher and by Anthony Parkhurst, including his reported discovery of copper in Newfoundland, Gilbert brought along a Saxon refiner who collected sample ores, purportedly of copper, iron, lead, and silver. See Carlos Slafter, *Sir Humphrey Gylberte and His Enterprise of Colonization in America* (Boston: Prince Society, 1903), 41.

10. Andrew Hadfield, *Amazons, Savages, and Machiavels. Travel and Colonial Writing in English, 1550–1630: An Anthology* (New York: Oxford University Press, 2001), 237.

11. Contemporaneous accounts, such as Hayes's, blamed his irascible temper, avarice, and poor judgment. Sir William Alexander, Knight, similarly identifies Gilbert's poor judgment, describing his desperation and "needlesse bravery," which led to his death; *An Encouragement to Colonies* (London, 1624), sig. D4 v–E r.

12. See Richard Hakluyt, *The Principal Navigations, Voyages, Traffiques, and Discoveries of the English Nation*, ed. Edmund Goldsmid (Edinburgh: E. & G. Goldsmid, 1887), 8:81–84, cited in Cell, *English Enterprise*, 43.

13. See Marc Lescarbot, *Noua Francia: or The Description of that Part of New France Which Is One Continent with Virginia* (London, 1609), sig. E3 r. The overwhelming effect of these sensations was sickness and disease, which was not aided by the seafaring conditions. "I would adde willingly to all the foresaid causes the bad foode of the sea, which in a long voyage brings much curroption in mans bodie. For one must of necessity, after foure or fie daies, luie of salt meate, or to bring sheepe aliue, and store of poultry . . . and the waters stinking" (sig E3 v).

14. As Henry Wadsworth Longfellow wrote in a poetic tribute, "By the Seaside," Gilbert's colonial endeavors ended tragically: "southward, for ever southward, / They drift through dark and day; / And like a dream, in the Gulf-stream, / Sinking, vanish all away"; *Poetical Works* (London: Houghton Mifflin, 1891), 46.

15. See Alan Taylor, *Writing Early American History* (Philadelphia: University of Pennsylvania Press, 2006), 108.

16. As Mary C. Fuller has argued, Newfoundland shaped how most Englishmen imagined "America" in the sixteenth century. Despite that Newfoundland "was probably England's earliest landfall, and certainly its earliest land claim in North America," England's lessons in colonization there remain "startlingly unremembered and unrecorded"; *Remembering the Early Modern Voyage: English Narratives in the Age of European Expansion* (Houndsmill, U.K.: Palgrave Macmillan, 2008), 118.

17. Henry S. Burrage, ed., *Early English and French Voyages, Chiefly from Hakluyt, 1534–1608* (New York: Charles Scribner's Sons, 1906), 228.

18. Virginia's potent scent is a trope echoed in later voyages. Strachey recorded that "before we come in sight of yt thirty leages, we smell a sweet savour as is usually from off Cape Vincent, the South Cape of Spayne, if the wynde come from the Shoare." Captain Devries is more specific about the origins of this scent, reporting in 1630 that "The 2nd December, threw the lead in 14 fathoms sandy bottom and smelt the land, which gave a sweet perfume as the wind came from the northwest, which blew off land and caused these sweet odors. This comes from the Indians setting fires at this time of year to the woods and thickets, in order to hunt, and the land is full of sweet smelling herbs as sassafras, which has a sweet smell. When the wind blows out of the north west and the smoke too is driven to sea, it happens that the land is smelt before it is seen"; quoted in Philip Bruce, *Economic History of Virginia in the Seventeenth Century* (New York: Macmillan, 1896), 88.

19. See Alden T. Vaughan, "Sir Walter Ralegh's Indian Interpreters, 1584–1618," *William & Mary Quarterly*, 3rd ser., 59 (2002): 341–76, 343–57.

20. Though Raleigh seems to have intended the natives to act as interpreters, learning English, Thomas Harriot, and perhaps also John White, studied Algonquin. See

David Quinn, *Explorers and Colonies: America, 1500–1625* (London: Hambledon, 1990), 243–44.

21. Peter Hoffer, *Sensory Worlds in Early America* (Baltimore, MD: Johns Hopkins University Press, 2003), 19.

22. See David B. Quinn and Alison M. Quinn, eds., *The English New England Voyages, 1602–1608* (London: Hakluyt Society, 1983), 31.

23. On the pox, and its association with exogamous, foreign diseases (particularly in the new world), see Jonathan Gil Harris, *Foreign Bodies and the Body Politic: Discourses of Social Pathology in Early Modern England* (New York: Cambridge University Press, 1998), 14.

24. All quotations are from *Ben Jonson: "Cynthia's Revels," "Poetaster," "Sejanus," "Eastward Ho,"* ed. C. H. Herford, Percy Simpson, Evelyn Simpson, and Evelyn Mary Spearing Simpson (New York: Oxford University Press, 1986).

25. Jordan Goodman argues that Galenists prescribed guaiacum and sassafras as a cure for syphilis, while Parcelsians prescribed mercury; see *Tobacco in History: The Cultures of Dependence* (London: Routledge, 1993), 43.

26. Once commonly used in the production of root beer soda and in tea, sassafras connotes a unique taste rather than scent. Sassafras is now banned by both the FDA and the IFA for use in either drinks or perfumes due to its carcinogenicity. Interestingly, it is surmised that native collection practices avoided the poisonous white-stemmed sassafras leaves, which are carcinogenic, collecting only red-stemmed leaves. See Maurice M. Iwu and Jacqueline C. Wooton, eds., *Ethnomedicine and Drug Discovery* (New York: Elsevier Science, 2002), 131.

27. For more on the ways in which the new world became a "wasteland," see Patricia Seed, *American Pentimento: The Invention of Indians and the Pursuit of Riches* (Minneapolis: University of Minnesota Press, 2001), chap. 2.

28. See Quinn and Quinn, *English New England Voyages*, 87.

29. See Donald Culross Peattie, *A Natural History of North American Trees* (New York: Houghton Mifflin, 2007), 276–77.

30. Francis Higginson, *New-Englands Plantation* (London, 1630), sig. B3 r. For example, orris root, a popular aromatic ingredient used to scent cloth, hair, and tooth powders, was difficult to cultivate commercially; it was routinely reported as common to the new world, despite the fact that it is not native to North America. Other aromatic ingredients thought to be discovered include benzoin (probably North American spicebush), cassia (perhaps dogwood, although *cassia* was often a generic term for aromatic wood), and china root (probably a smilax). Such confusions suggest that smell was an inexact sensory guide in unfamiliar environments. Discovery of new kinds of aromatic ingredients, such as sweet flag, or *Acorocus calamus*, and sarsaparilla, resulted from contact with natives. Of the numerous aromatic properties discovered in the new world, sassafras was the most prominent and the most illusive; it was also one of the few native medicines that explorers and colonists preferred to their own. See Charlotte Erichson-Brown, *Medicinal and Other Uses of North American Plants: A Historical Survey with Special Reference to the Eastern Indian Tribes* (Mineola, NY: Dover Publications, 1989), xi.

31. For example, Master Robert Meriton in the Archer expedition, Robert Salterne and Thomas Bridges in the Pring expedition, and James Roseier in the Waymouth expedition are listed as identifying sassafras (and other new world plants), yet no explanation of their expertise is listed, though perhaps previous experience in southern trips aided in the discovery of this difficult specimen. See Quinn and Quinn, *English New England Voyages*, 16. *Sagacity*, Susan Scott Parish argues, was the word most often used by English naturalists to describe their native informants, a term that emphasized their "animal acumen," particularly their sense of smell; *American Curiosity: Cultures of Natural History in the Colonial British Atlantic World* (Chapel Hill: University of North Carolina Press, 2005), 239.

32. Goodman, *Tobacco in History*, 44.

33. For more on tobacco, see Marcy Norton's *Sacred Gifts, Profane Pleasures: A History of Chocolate and Tobacco in the Atlantic World* (Ithaca, NY: Cornell University Press, 2008), which examines both pre- and post-Hispanic culture in the new world; Jordan Goodman's *Tobacco in History*, which examines the commodity history of tobacco in Enlightenment Europe; and Jeffrey Knapp's "Elizabethan Tobacco" *Representations* 21 (1988): 26–66. One exception is Charles Manning and Merrill Moore's "Sassafras and Syphilis," in *New England Quarterly* 9.3 (1936): 473–75.

34. John Frampton, *Ioyfull Newes out of the Newe Founde Worlde . . .* (London, 1577), sig. Mij r.

35. Ibid.

36. See Donald Beecher, "John Frampton of Bristol, Trader and Translator," in *Travel and Translation in the Early Modern Period*, ed. Carmine G. Di Biase (New York: Rodopi, 2006), 103–122, 103.

37. Frampton was connected with influential Bristol merchants, many of whom founded the Spanish Company in 1577. All six of Frampton's influential translations deal with trade, including two navigational texts, three on China, and one on Portuguese trade.

38. See Frampton, *Ioyfull Newes*, sig. Mij r.

39. Ibid., sig. Mii r, Nii r, Nii v, Niv r.

40. Ibid., sig. Mij r.

41. Ibid.

42. Ibid., sig. Nii r.

43. Ibid. Lawrence C. Wroth describes Monardes as "an eager investigator haunting the Seville docks in order to meet the ships which would bring him new plants or new regimes for those already growing in his garden"; "An Elizabethan Merchant and Man of Letters," *The Huntington Library Quarterly* 17:4 (August 1954): 299–314, 306.

44. See William C. Sturtevant and David B. Quinn, "This New Prey: Eskimos in Europe in 1567, 1576, and 1577," *Indians and Europe: An Interdisciplinary Collection of Essays*, ed. Christian Feest (Lincoln: University of Nebraska Press, 1999), 61–140, 103.

45. See Paul Hulton, *The Work of Jacques Le Moyne De Morgues: A Huguenot Artist in France, Florida, and England*, 2 vols. (London: British Museum Publications Limited, 1977), 1:118.

46. The first depiction of Ribault's marker emphasizes this point. The marker bears only the French arms, depicted as a tiny square, with an even tinier escutcheon. See plate 104 in Hulton, *Work of Jacques Le Moyne De Morgues*, 1:98, also 2:204.

47. Ibid., 1:141.

48. Ibid.

49. Christian Feest argues that this raises serious doubts about Le Moyne's attribution as painter of these images; "Jacques Le Moyne Minus Four," *European Review of Native Studies* 1.1 (1988): 33–38.

50. Thomas Harriot, *A Briefe and True Report of the New Found Land of Virginia* [1590], ed. Paul Hulton (New York: Dover, 1972), sig. A2 r.

51. Harriot writes: "Most thinges they sawe with vs, as Mathematicall instruments, sea compasses, the vertue of the loadstone in drawing yron, a perspectiue glasse whereby was shewed manie strange sightes, burning glasses, wildefire woorkes, gunnes, bookes, writing and reading, spring clocks that seeme to goe of themselues, and manie other thinges that wee had, were so straunge vnto them, and so farre exceeded their capacities to comprehend the reason and meanes how they should be made and done, that they thought they were rather the works of gods then of men, or at the leastwise they had bin giuen and taught vs of the gods"; ibid., sig. C4 r.

52. Ibid., sig. D r.

53. Taylor, *Writing Early American History*, 23.

54. Karen Ordahl Kupperman, *Indians and English: Facing Off in Early America* (Ithaca, NY: Cornell University Press, 2000), 178.

55. Harriot, *Briefe and True Report*, sig. C4 v.

56. Ibid.

57. Ibid., sig. B r.

58. Other aromatic ingredients mistakenly reported as being discovered in the new world include benzoin (probably North American spicebush), cassia (perhaps dogwood, although *cassia* was often a generic term for aromatic wood), and china root (probably a smilax). Such confusions suggest that smell was an inexact sensory guide in unfamiliar environments.

59. Ibid., sig. B v.

60. Harriot, *Briefe and True Report*, sig. B2 r.

61. See Ute Kuhlemann, "Between Reproduction, Invention, and Propaganda: Theodor de Bry's Engravings after John White's Watercolours," in Sloan, *New World*, 79.

62. Sloan, *New World*, 6.

63. Joyce Chaplin argues that John White's watercolors are a theatrical view of the new world. "They dance. They wave to each other. They prepare food. They smile. . . . In a very real sense, White introduced Virginia's natives to the English as if he were displaying them in a theatre"; "Roanoke 'Counterfeited' According to the Truth," in Sloan, *New World*, 51, 59.

64. Hakluyt, *Principal Navigations*, vol. 8 (New York: Macmillan, 1906), 406.

65. Oxford English Dictionary, 2nd ed., s.v. "noisome," n. 4. For more, see Holly Dugan, "Coriolanus and 'the Rank-Scented Mienie': Smelling Rank in Early Modern Lon-

don," in *Masculinity and the Metropolis of Vice, 1550–1750,* ed. Amanda Bailey and Roze Hentschell (Houndsmill, U.K.: Palgrave Macmillan, 2010), 139–59.

66. John White, "To Master Richard Hakluyt, written from my house at Newton, in Kilmore, the 4th of February, 1593," in *History of North Carolina: With Maps and Illustrations,* ed. Francis Lister Hawks (Fayetteville, NC: E. J. Hale & Son, 1857), 214–15, 215.

67. Harriot, *Briefe and True Report,* sig. C2 v.

68. Ibid.

69. "This Vppówoc is of so precious estimation amongst then, that they thinke their gods are maruelously delighted therwith: Whereupon sometime they make hallowed fires & cast some of the pouder therein for a sacrifice . . . also after an escape of danger, they cast some into the aire likewise: but all done with strange gestures, stamping, sometime dauncing, clapping of hands, holding up of hands, & staring up into the heauens, uttering therewithal and chattering words & noises"; ibid., sig. B4 v.

70. See Hoffer, *Sensory Worlds,* 29.

71. See Quinn and Quinn, *English New England Voyages,* 206. Though it is impossible to say for certain how much an early modern "ton" weighed, Raleigh's reference to the *Concord*'s cargo as both a "tunne" and as "2200" pounds comport to a "long ton," a common weight of measurement used in early modern shipping pounds to connote 2,240 pounds. See Frederic C. Lane, "Tonnages, Medieval and Modern," *Economic History Review,* 2nd ser., 17.2 (1964): 213–33.

72. In 1602, sassafras was priced at about 20 shillings a pound, until its price dropped rapidly. Raleigh, investigating the drop, discovered reports of sassafras sales in England's southern ports, all from a recent voyage to northern Virginia.

73. See Quinn and Quinn, *English New England Voyages,* 204.

74. In a letter to Cecil dated August 21, 1602, Raleigh charges that Gilbert has destroyed the sassafras market: "whereas sarsephraze was worth 10s, 12s, & 20s apound before Gilbert returned his cloying of the market will overthrow all myne & his also / he is contented to have all stayed, not only for this present, butt being to go agayne others will also go & destroy the trade, which otherwise would yield 8, or 10, for on certenty, & a returne in xx weekes"; Quinn and Quinn, *English New England Voyages,* 206.

75. Gosnold was the captain and Brereton, the shipmaster.

76. Quinn and Quinn, *English New England Voyages,* 206.

77. Brereton is explicit in this accusation: "When they came along the coast to seeke the people, they did not pretending that the extremitie of weather and losse of some principall ground-tackle, forced and feared them from searching the port of Hatarask, to which they were sent. From that place where they abode, they brought aboard Sassafras, Radix Chinae or the Chinae root, Beniamin, Cassia lignea, a rinde of a tree more strong than any spice as yet knowen [most likely sweet bay]"; M. John Brereton, *A Briefe and True Relation of the Discouerie of the North Part of Virginia, Being a Most Pleasant, Fruitfull and Commodious Soile,* (London, 1602), sig. B4 v.

78. Quinn and Quinn, *English New England Voyages,* 115–16.

79. The account makes this connection explicitly: "[W]ith a piece of Chalke described the Coast there abouts, and could name Placentia of the New-found-land, they spake div-

ers Christian words and seemed to understand much more than we, for want of Language could comprehend." Archer focuses on visual recognition through clothing: "One that seemed to be their Commander wore a Wastecoate of black worke, a paire of breeches, cloth Stockings, Shooes, Hat, and Band, one of the two more had also a few things made by some Christians." Some scholars interpret the "Placentia of the New-found-land" to be a Basque reference. Quinn and Quinn, *English New England Voyages*, 117.

80. Brereton, *Briefe and True Relation*, sig. B2 r–v.

81. Ibid., sig. B v.

82. Ibid., sig. A4 r.

83. Ibid., sig. B3 r.

84. After describing the fair conditioned people and the activities of trade, Archer again records that on "the first of June, we employed our selves in getting Sassafrage and the building of our Fort. . . . [O]n the fifth we continued our labor . . . the company with me only eight persons." Approached by fifty "Indians in hasite manner," among them a leader of "authority," Archer "mooued my selfe towards him seuen or eight steps, and clapt my hands first on the sides of mine head, then on my breast, and after presented my Musket with a threatening countenance, thereby to signifie unto them I either a choice of Peace or Warre." According to Archer, such motions worked: "[H]ee vsing me with mine owne signes of Peace, I stept forth and imbraced him, his company then all sate downe in manner like Grey-hounds upon their heeles with whom my company fell a bartering." Here, the narrative emphasizes their "maruelling" at "bright and sharpe" knives. Quinn and Quinn, *English New England Voyages*, 134.

85. The crises avoided, the natives then labor with the English: "Our dinner ended, the Seignior first took leaue and departed, next all the rest sauing foure that stayed and went into the Wood to helpe vs digge Sassafrage, whom we desired to goe aboord vs, which they refused and departed." Archer does not comment on the natives' reluctance to board the native ship, nor their motives in extending the offer. He only continues to chronicle the labor of collecting sassafras: "the fourteenth, fifteenth, and sixteent wee spent in getting Sasafrage and fire-wood of Cedar, . . . the seuenteenth, we set sayle"; ibid., 133–38.

86. Not surprisingly, when he returned to England, Gosnold corroborated Raleigh's assessment of England's sassafras market. In a letter to his father dated September 7, 1602, Gosnold describes that sassafras was easily procured from Elizabeth's island, with "little disturbance" was "reasonable plenty," but he reports that the bulk of their labor was not spent on collecting sassafras since "we were informed before our going forth that a tunne was sufficient to cloy England"; ibid., 210. Gosnold's assessment of England's sassafras market sounds suspiciously like Raleigh's; the *Concord*, after all, brought back just short of a long ton.

87. In Martin Pring's narrative, smell provides a proximate and shared experience of the environment, which sometimes facilitates trade and sometimes facilitates violence. For example, although the sailors were struck by the size of native boats, it was the sweet-scented Rozen that captivated their narrative description: "Their Boats . . . were in proportion like a Wherrie of the Riuer of Thames, seventeen foot long and four foot broad, made of the Barke of a Birch-tree, farre exceeding in bignesse those of

England: it was sowed together with strong and tough Oziers or twigs, and the seames couered over with Rozen or Turpentine little inferiour in sweetness to Frankincense"; ibid., 223.

88. After loading a small barke with "as much Sassafras as we thought sufficient, and sent her home to England," the remaining crew renewed their efforts at collecting it, continuing to load the remaining ship with its cargo. The natives surrounded the napping Englishmen, who were armed with muskets and two English mastiffs. Pring, on waking and discovering their predicament, fired a loud warning shot from a "piece of great Ordnance," in order "to giue terrour to the Indians, and warning to our men which were fast asleepe in the Woods." Yet Pring's defense did not work. It took a subsequent shot to rouse both the men and the dogs. Only then did they jump to attention. When the mastiff, "with an halfe Pike in his mouth," charged the native boat, the natives "turned all to a iest and sport, and dparted away in a friendly manner"; ibid., 227.

89. Michael Drayton, "To the Virginian Voyage," *Poems* (London, 1620), sig. Pp 5 r–v.

90. Hoffer, *Sensory Worlds*, 19.

91. Ibid., 49.

92. Ibid., 72.

93. Philip Boucher makes this point about sassafras in relationship to the Huguenots' presence in Florida; "Revisioning the 'French Atlantic': or, How to Think about the French Presence in the Atlantic, 1550–1625," in *The Atlantic World and Virginia*, ed. Peter C. Mancall (Chapel Hill: University of North Carolina Press, 2007), 274–306.

94. Philip Alexander Bruce, *Economic History of Virginia in the Seventeenth Century: An Inquiry into the Material Condition of the People, Based upon Original and Contemporaneous Records* (New York: Macmillan, 1907), 1:92–93.

95. William Wallace Tooker, "Some More about Virginia Names," *American Anthropologist* 7 (1910): 524–28.

96. Hoffer, *Sensory Worlds*, 72.

97. John Smith, "A True Relation" [1608] in Philip Barbour, ed., *Jamestown Voyages Under the First Charter: 1606–1609* (New York: Cambridge University Press, 1969), 1:170, 172.

98. John Lawson, *A New Voyage to Carolina* [1709], ed. Hugh Talmage Lefler (Chapel Hill: University of North Carolina Press, 1967), cited in Joyce Chaplin, *Subject Matter: Technology, the Body, and Science on the Anglo-American Frontier, 1500–1676* (Cambridge, MA: Harvard University Press, 2001), 3.

99. Chaplin, *Subject Matter*, 3.

100. Jean Jacquot, "Thomas Harriot's Reputation for Impiety," *Notes and Records of the Royal Society* 9 (1952): 164–87, cited in ibid., 31.

Chapter 4 · Smelling Disease

1. Thomas Dekker, *The Wonderfull Yeare: Wherein Is Shewed the Picture of London, Lying Sicke of the Plague* (London, 1603).

2. Ibid., sig. C3 v.

3. Ibid.

4. Winds from the south and west were especially feared, particularly in places northeast of London. See Paul Slack, *The Impact of the Plague in Tudor and Stuart England* (London: Routledge & Keegan Paul, 1985), 30.

5. See Jonathan Gil Harris, *Sick Economies: Drama, Mercantilism, and Disease in Shakespeare's England* (Philadelphia: University of Pennsylvania Press, 2003), 113.

6. S.H., *A New Treatise of the Pestilence* (London, 1603), sig. A2, cited in Harris, *Sick Economies*, 113.

7. Slack, *Impact of the Plague*, 155.

8. John Nichols, ed., *The Progresses and Public Processions of King James the First* (London: J. B. Nichols, 1828), 1:198–99.

9. Dekker, *The Wonderfull Yeare*, sig. C3 v.

10. Ibid., sig. D3.

11. Ibid., sig. B r–v.

12. Ibid., sig. B v.

13. Ibid., sig. C r–v.

14. Edward Hall, *The Union of the Two Noble and Illustrate Famelies of Lancastre [and] Yorke* (London, 1548), fol. clxvi, sig. ccciiii r.

15. Pamela Nightingale, *A Medieval Mercantile Community: The Grocers' Company and the Politics and Trade of London, 1000–1485* (New Haven: Yale University Press, 1995), 81.

16. Thomas Dekker, *The Magnificent Entertainment Given to King James, Queen Anne, His Wife, and Henry Frederick the Prince . . .* (London, 1604), n.p.

17. The link between odor and Henry's rousing speech might have been implicitly strengthened by the early modern cultural association of this feast day with the leather industry as St. Crispin was the patron saint of the shoemaker's guild. See Alison A. Chapman, "Whose Saint Crispin's Day Is It?: Shoemaking, Holiday Making, and the Politics of Memory in Early Modern England," *Renaissance Quarterly* 54 (2001): 1467–94.

18. All quotations from *Henry V* are from *The Norton Shakespeare: Based on the Oxford Edition*, 2nd ed., ed. Stephen Greenblatt, Walter Cohen, Jean E. Howard, and Katharine Eisaman Maus (New York: W.W. Norton, 2008), 1471–1548.

19. Thomas Lodge, *A Treatise of the Plague* (London, 1603), sig. B2 r.

20. Stephen Bradwell, *A Watch-man for the Pest . . .* (London, 1625), preface, sig. A2 r–v.

21. All quotations from *Romeo and Juliet* are from Greenblatt et al., *Norton Shakespeare*, 905–72.

22. Dekker, *The Wonderfull Yeare*, sig. C3.

23. All citations are from *A Faire Quarrell* in *Thomas Middleton: The Collected Works*, ed. Gary Taylor et al. (New York: Oxford University Press, 2008), 1209–51.

24. Alfred Plummer, *The London Weavers' Company, 1600–1970* (New York: Routledge, 1972), 186. Slack, *Impact of the Plague*, 154. During the 1665 outbreak of the plague, a giant pit was dug in the cemetery and thousands of bodies were buried there. See Vanessa Harding, *The Dead and the Living in Paris and London, 1500–1670* (New York: Cambridge University Press, 2002), 66.

25. Michel Berlin, "Reordering Rituals: Ceremony and the Parish, 1520–1640," in *Londinopolis: Essays in the Cultural and Social History of Early Modern London*, ed. Mark S.

Jenner and Paul Griffiths (Manchester, U.K.: Manchester University Press, 2002), 47–66, 57.

26. All quotations from *Cymbeline* are from Greenblatt et al., *Norton Shakespeare,* 2963–3054.

27. John Davies, *Humours Heau'n on Earth; with . . . The Triumph of Death: or, The Picture of the Plague, According to the Life; as it was in Anno Domini 1603* (London, 1609), sig. K2.

28. See F. W. Brownlow, *Shakespeare, Harsnett, and the Devils of Denham* (Wilmington: University of Delaware Press, 1993), 85.

29. Samuel Harsnett, *A Declaration of Egregious Popish Impostures* (London, 1603), sig. G r.

30. Ibid., sig. S r.

31. Thomas Wright, *Passions of the Minde in Generall,* ed. Thomas Sloane (Urbana-Champagne: University of Illinois Press, 1971), 177.

32. See Richard Brome, *The Antipodes* (London, 1640), sig. K v, 5.1. I am grateful to Keith Botehlo for alerting me to this exchange. For more on sensory witnessing, see his *Renaissance Earwitnesses: Rumor and Early Modern Masculinity* (Houndsmill, U.K.: Palgrave Macmillan, 2008).

33. See Mary Floyd-Wilson, *English, Ethnicity, and Race in Early Modern Drama* (New York: Cambridge University Press, 2005).

34. Consider, for example, the widespread belief in–and access to–herbs and objects used as aphrodisiacs in the period. See Sir Kenelm Digby, *Choice and Experimented Receipts in Physick and Chirurgery, as Also Cordial and Distilled Waters and Spirits, Perfumes, and Other Curiousities* (London, 1668), 113, 30, 40. Similarly, Sir James Harrington is troubled by the holy implications of airborne influences. In a 1669 religious meditation titled *A Holy Oyl and a Sweet Perfume,* Harrington muses that smell's invisible nature forces him to rely on imperfect bodily effects as a gauge: "[I]f venomous and malignant smells, and spirations, have demonstrably, and undeniably a secret and virulent Power, to destroy life and nature; Then assuredly, by the rule of contraries, and according to right reason, redolant, fragrant, cordial, and spiritual odours, scents, aires, and smells, do not only refresh and exhilerate, but fortifie, preserve and nourish an life and beings"; *A Holy Oyl; and, a Sweet Perfume: Taken out of the Sanctuary of the Most Sacred Scriptures . . . Poured Forth into Eight [or Rather Five] Vessels* (London, 1609), sig. Hhh1 v–Hhh2 r.

35. Carol Rawcliffe, *Medicine and Society in Later Medieval England* (Stroud, U.K.: A. Sutton, 1995), 42.

36. Ann Rosalind Jones and Peter Stallybrass, *Renaissance Clothing and the Materials of Memory* (New York: Cambridge University Press, 2001), 204.

37. See ibid., 34, 46, 50.

38. See Gail Kern Paster, "Purgation as the Allure of Mastery: Early Modern Medicine and the Technology of Self," *Material London, ca. 1600,* ed. Lena Cowen Orlin (Philadelphia: University of Pennsylvania Press, 2000), 193–206; Michael Schoenfeldt, *Bodies and Selves in Early Modern England: Physiology and Inwardness in Spenser, Shakespeare, Herbert, and Milton* (New York: Cambridge University Press, 1999); Harris, *Sick Econo-*

mies; Mary J. Dobson, *Contours of Death and Disease in Early Modern England* (New York: Cambridge University Press, 2003); and Thomas Lacquer, *Making Sex: Body and Gender from the Greeks to Freud* (Cambridge, MA: Harvard University Press, 1990).

39. Kathleen Brown, *Foul Bodies: Cleanliness in Early America* (New Haven, CT: Yale University Press, 2009), 41; Schoenfeldt, *Bodies and Selves,* 106.

40. Literary scholars and historians have expounded the scientific and literary links between humoral theory and cultural representation, especially in exploring the early modern histories of gender and sexuality. Some critics focus the structures of discipline and shame engendered by humoral theory; others "stress empowerment that Galenic physiology and ethics bestowed upon the individual"; Schoenfeldt, *Bodies and Selves,* 11. For an analysis of how humoral theory shaped cultural representations, especially constructions of race, gender, and sexuality, see Joyce Chaplin, *Subject Matter: Technology, the Body, and Science on the Anglo-American Frontier* (Cambridge, MA: Harvard University Press, 2001), 122, and Floyd-Wilson, *English Ethnicity,* 26.

41. See John Leeds Barroll, *Politics, Plague, and Shakespeare's Theater: The Stuart Years* (Ithaca, NY: Cornell University Press, 1991), chap. 3.

42. Helkiah Crooke, Caspar Bauhin, and André Du Laurens, *Mikrokosmographia. A Description of the Body of Man* (London, 1615), 619. For Crooke, the nose is the vehicle that draws air into the body, while the mind interprets the odor: "The body therefore that perceiueth or apprehendeth odours is placed higher . . . It remayneth therefore that it must bee those processes which because they are somewhat like the nipples of a Dugge are called Mammillares" (ibid.). Olfaction blurs the line between the "perceiving body" of man and his more animal instincts.

43. "There is no perception of odours, except when we do inspire, for though you fill the Nose full with Muske or Ambergreese or other odoriferous bodyes; yea though you should annoynt the whole membrame with sweet oyles, yet you shall haue no perception of odours except you draw in the Ayre by inspiration"; ibid., 712.

44. Ibid., 647.

45. Ibid., 705. Modern scientists preserve this concept: smell is the only unmediated sense, with external stimuli directly interacting with the brain's receptors. See Jelena Djordejvic and Marilyn Jones-Gotman, "Olfaction and the Temporal Lobes," in *Olfaction and the Brain,* ed. Warrick Brewer et al. (New York: Cambridge University Press, 2006), 28–49.

46. Crooke, *Mikrokosmographia,* 704–5.

47. Ibid., 705.

48. Ibid., 251.

49. Ibid., 251, 252.

50. Ibid., 251, 252. Tallow candles were widely used in this period and emitted a strong stench.

51. Indeed, Crooke reports that Placentinus's experiments prove this fact: "But it may bee objected, that if we hold our breath wee cannot smell, and therefore this Sense is not accomplished without Inspiration. Placentinus answereth that it is not true, that if an odour bee applied to the Nose and the breath retained, no Sense will bee made"; ibid., 701.

52. Thomas Thayre, *An Excellent and Best Approved Treatise of the Plague . . .* (London, 1625), 13.

53. Crooke, *Mikrokosmographia,* 711.

54. Ibid., 710.

55. Ibid., 713.

56. Ibid., 712.

57. Air from animals was thought to linger longer than plant-based air. See Sloane MS 1821, British Library, London. See also Dobson, *Contours of Death,* 42.

58. Gail Kern Paster, *The Body Embarassed: Drama and the Disciplines of Shame in Early Modern England* (Ithaca, NY: Cornell University Press, 1993), 11.

59. Renaissance medical thinkers identified the plague as resulting from corrupt air. See Ambroise Paré, *A Treatise of the Plague* (London, 1630), 4–5. On Hippocratic definitions of health, defined as the connection between air, water, and soil, see Dobson, *Contours of Death,* 10, and Floyd-Wilson, *English Ethnicity,* 23.

60. Miasmas were poisonous particles accompanied by bad smells. See Harris, *Sick Economies,* 113.

61. Margaret J. Healy, *Fictions of Disease in Early Modern England* (Houndsmill, U.K.: Palgrave Macmillan, 2001), 54.

62. Dobson, *Contours of Death,* 485.

63. Ibid., 9, 225, 450, 85, 26.

64. Healy, *Fictions of Disease,* 38, 113.

65. Elizabeth I, *Orders, Thought Meete by Her Majesties and Her Privie Counsell, to Be Executed Throughout the Counties of This Realme, in Such Townes, Villages, and Other Places, as Are, or May Be Hereafter Infected with the Plague* (London, 1593); James I, *Orders Thought Meete by His Majestie, and His Privie Counsell, to Be Executed Throughout the Counties of This Realme, in Such Townes, Villages, and Other Places, as Are, or May Be, Hereafter Infected with the Plague* (London, 1603).

66. See Sloane MS 1471, fol. 18–20; see also Bradwell, *Watch-man for the Pest,* sig. A2 v.

67. See Rachel Munkhoff, "Searchers of the Dead: Authority, Marginality, and Interpretation of Plague in England," *Gender & History* 11.1 (1999): 1–29, 24 n. 5.

68. Scents, trapped in clothing, were a particular danger. When someone died from the plague, all bed linens and clothing of the diseased were burned. For a brief discussion of the material practices that coded wardens' and searchers' contact with the infected, see Charles F. Mullett, "Some Neglected Aspects of Plague Medicine in Sixteenth-Century England," *The Scientific Monthly* 4.4 (1937): 3325–37, 3326.

69. See Healy, *Fictions of Disease,* 52. Epidemiologists, however, disagree on interpretations of mortality rates, arguing that incubation periods varied widely depending on various strains of plagues; Dobson, *Contours of Death,* 483.

70. Plague quarantines lasted between twenty-eight and forty days. The term "quarantine" marked a forty-day period, derived from legal discourse as the amount of time a widow, entitled to a dower, was allowed to remain in her deceased husband's house. Later in the seventeenth century, this term applied to a forty-day period in which a ship suspected of carrying contagions was required to remain offshore. See Mullett, "Some Ne-

glected Aspects of Plague Medicine," 326; Dobson, *Contours of Death,* 486; and Munkhoff, "Searchers of the Dead," 10.

71. See Elizabeth I, *Orders,* and James I, *Orders.* Daniel Defoe's eighteenth-century memoir of the 1665 plague also remarks on the impact of the plague on churches:

"Once, on a public day, whether a Sabbath-day or not I do not remember, in Aldgate Church, in a pew full of people, on a sudden one fancied she smelt an ill smell. Immediately she fancies the plague was in the pew, whispers her notion or suspicion to the next, then rises and goes out of the pew. It immediately took with the next, and so to them all; and every one of them, and of the two or three adjoining pews, got up and went out of the church, nobody knowing what it was offended them, or from whom. This immediately filled everybody's mouths with one preparation or other, such as the old woman directed, and some perhaps as physicians directed, in order to prevent infection by the breath of others; insomuch that if we came to go into a church when it was anything full of people, there would be such a mixture of smells at the entrance that it was much more strong, though perhaps not so wholesome, than if you were going into an apothecary's or druggist's shop. In a word, the whole church was like a smelling-bottle; in one corner it was all perfumes; in another, aromatics, balsamics, and variety of drugs and herbs; in another, salts and spirits, as every one was furnished for their own preservation." Daniel Defoe, *A Journal of the Plague Year* (New York: Modern Library Classic, 2001), 198.

72. Crooke, *Mikrokosmographia,* 703.

73. Anonymous, *Appendix of Venereal Disease in Men and Women,* 161, in Sloane MS 1821.

74. Ibid.

75. Thayre, *Treatise of the Plague,* sig. 1B r.

76. Epidemiologists suggest that flea-infected summer months could account for these statistics. See Dobson, *Contours of Death,* 483.

77. For plague related deaths in London in 1603, see James Godskall, *The Arke of Noah, for the Londoners That Remaine in the Cittie to Enter in, with Their Families, to be Preserved from the Deluge of the Plague* (London, 1604). For an account of the plague of 1625, see Sloane MS 1471. For the plague of 1665, see William Austin, *Epiloimia Epe, or, the Anatomy of the Pestilence* (London, 1662), and MS 7382/3, Wellcome Library, London. For a relationship between smells and heat, see Joannes de Mediolano, *The Englishman's Doctor* (London, 1607): "Though all ill savours do not breed infection, / Yet sure infection commeth most by smelling / Who smelleth still perfumed, his complexion / Is not perfum'd by Poet Martials telling, / Yet for your lodging roomes give this direction, / In houses where you mind to make your dwelling, / That nere the same there be no evill sents / Of Puddle-waters, or of excrements, / Let ayre be cleare and light, & free from faults, / That come of secret passages and vaults"; sig. A9 r.

78. Thayre, *Treatise of the Plague,* 8.

79. Ibid.

80. See Elizabeth I, *Orders,* and James I, *Orders.*

81. Although published in the eighteenth century, Defoe's *Journal of the Plague Year* provides a fascinating take on the role of perfumed air in London's 1665 plague; see pp. 41, 75, 230.

82. MS 809, Wellcome Library, London; Anonymous, *Present Remedies against the Plague* (London, 1592); W.M., *The Queen's Closet Opened . . .* (London, 1655), 8, 39, 40; Anonymous, *The Treasurie of Hidden Secrets* (London, 1600), chap. 51.

83. Continental disease control included the development of pest houses, or Lazar houses, as well as this type of protective clothing. Some historians surmise that the erratic pattern of contagion in London and other English cities made continental models useless. For a discussion of plague beaks, see Christine M. Boeckl, *Images of Plagues and Pestilence: Iconography and Iconology* (Kirksville, MO: Truman State University Press, 2000). For a discussion of how London differed from continental European cities, see Dobson, *Contours of Death*, 26.

84. Most wealthy professionals, including doctors, fled London during plague years. For a woodcut depicting their return, see Anonymous, "London Welcomes Home Her Runaways," *The Countrie Ague* (1625), cited in Frank Wilson, *The Plague in Shakespeare's London* (Oxford: Claredon Press, 1923), 163. Not everyone could flee the city in times of trouble. For a moving early modern account pondering the economic impact and risk of such a decision, see Wellcome MS 7382/3.

85. See Elizabeth I, *Orders*, and James I, *Orders*.

86. Edwards, *A Treatise Concerning the Plague and Pox* (London, 1652), sig. Ggg v.

87. See Healy, *Fictions of Disease*, 61. For a representative example: "CHAP. XII. Doth shew what you must do when you go to visit the sick. First before you enter into the house, command that a great fire be made in the chamber where the sick lieth, and that some odoriferous perfume be burnt in the midst of the chamber"; Edwards, *Treatise Concerning the Plague*, sig. Ggg v–Ggg ii r.

88. The first European pomander appeared in an inventory in the thirteenth century. By the fourteenth century, small metal containers holding scented pastes or sponges were routinely used; by the sixteenth century, pomanders had evolved to include hinged devices, attached to chains, rings, and clothing. See Annette Green and Linda Dyett, *Secrets of Aromatic Jewelry* (Paris: Flammarion, 1998), 49–50.

89. Thomas Lodge's 1523 recipe for a pomander requires only fruit, wax, and strong aromatics: "To make a Pomander or good to preserue the bearers thereof from the plague: Take the quantity of a good Apple of yellow waxe, a good sponnefull of tarr and 5 or 6 sponnefulls of good wine vinegar boyle these together and then put thereto so much of the poweder of wormwood as will make it ery thicke and when it is well stirred together, take it off the fier, and when it is cold make it up in balles and make a hole thorow out of them, and with a stringe weare it about your necks in tyme of sickness and by good grace it will preserve you from infection also a peece of the roote of any olio or a peece of the pile of a lemmon or orange, or a leafe of sorrill any of these being carried in your mouth and thereof a little is very good against the infection"; Additional MS 34212, British Library, London, fol. 95.

90. Many "pomanders" of prayers were published in the sixteenth century, including Thomas Becon, *Pomander of Prayer* (London, 1563), reprinted in 1565 and 1578; Richard

Whitford, *Pomander of Prayer* (London, 1531); and Richard Whitford, *Pomander of Prayer* (London, 1532).

91. For a rare example of an inscribed pomander, see "Silver Gilt Pomander in the Form of a Snail," Museum of Science and Industry, London, A642178.

92. Anonymous, *Treasurie of Hidden Secrets* (London, 1600); "Receipt Book of Mary Dogget," Additional MS 27466, and "Poor Man's Talent" Additional MS 34212, British Museum, London; MS 182, Wellcome Library, London; Thayre, *Treatise of the Plague;* Philbert Guibert, *The Charitable Physitian* . . . (London, 1639); Edwards, *Treatise Concerning the Plague;* W.M., *A Queen's Delight;* G. Markham, *The English House-wife* . . . (London, 1683).

93. See for example, "Pestilencial Antidotes," in Anonymous, *Appendix of Venereal Disease,* in Sloane MS 1821, fol. 215.

94. Thayre, *Treatise of the Plague,* 13.

95. Philip Stubbes, *The Anatomie of Abuses* . . . (London, 1583), sig. G r.

96. Brathwaite continues, seemingly defending the sense: "[S]o it does: but what especiall delights confers it for one of these inconveniences; cheering the whole bodie with the sweetest odours, giuing libertie to the vitall powers, which otherwise would be imprisone." Two lines later, he abruptly concludes: "This *Sence* of mine shall not be subiected to outward delicacies: Let the Courtier *smell* of perfumes, the sleeke-fac'd Lady of her paintings, I will follow the *smell* of my Sauiours oyntments: how should I be induced, following the direction of reason, by such soule-bewitching vanities, which rather peruert the refined lustre of the minde, than adde the least of perfection to so excellent an essence?"; Richard Brathwaite, *Essaies upon the Five Senses with a Pithie One upon Detraction* . . . (London, 1620), 58–59.

97. Michael Snodin and John Styles, *Design and the Decorative Arts: Britain 1500–1900* (London: Victoria and Albert Museum, 2001), 110.

98. John Dobson and Adrian Scoop's Commonplace book, with an account of the 1625 plague, Sloane MS 1471.

99. By the seventeenth century, early modern medical authorities had separated the spread of poxes (syphilis) from pestilences (plagues). However, the link between the miasmic theory of contagion and the pox lingered in early modern popular imagination; see Harris, *Sick Economies,* 96. For an extended reading of perfume in relationship to sodomy, see Richard Halpern, *Shakespeare's Perfume: Sodomy and Sublimity in the Sonnets, Freud, Wilde, and Lacan* (Philadelphia: University of Pennsylvania Press, 2002).

100. Stubbes, *Anatomie of Abuses,* sig. Fix v.

101. Edmund Spenser, *Amoretti and Epithalamion* (London, 1595), sig. E v.

102. For a discussion of how the blazon functions within Renaissance poetry, see Nancy J. Vickers, "Diana Described: Scattered Women and Scattered Rhyme," *Critical Inquiry* 8.2 (1981): 265–79.

103. Spenser, *Amoretti and Epithalamion,* sig. E v.

104. Ibid.

105. Robert Herrick, *Hesperides, or, the Workes Both Humane and Divine of Robert Herrick, Esq.* (London, 1648), 11.

106. The Oxford English Dictionary defines *redolent* as "having or diffusing a pleasant odor; sweet smelling, fragrant, odorous." In Herrick's osmologies, his lover's body is redolent; see Herrick, *Hesperides*, 11, 64.

107. Herrick notes this danger as well but prefers women's bodily odors to perfumes, which he believes mask dangers. His "On a Perfum'd Lady" asks: "You say y'are sweet; how sho'd we know / Whether that you be sweet or no? / From *Powders* and *Perfumes* keep free; / Then we shall smell how sweet you be"; ibid., title page, 11.

108. In Ben Jonson's *The Alchemist,* his titular character, Subtle, traverses quarantined spaces, dispensing pomanders to city wives: "Methinks I see him ent'ring ordinaries, / Dispensing for the pox, and plaguy houses, / Reaching his dose, walking Moorfields for lepers, / And off'ring citizens' wives pomander-bracelets"(1.4.1).

109. For an analysis of women's inheritances and moveable property, see Natasha Korda, *Shakespeare's Domestic Economies: Gender and Property in Early Modern England* (Philadelphia: University of Pennsylvania Press, 2002).

110. Jones and Stallybrass, *Renaissance Clothing,* 213.

111. Mary Monica Maxwell-Scott and Dominique Bourgoing, *The Tragedy of Fotheringay, Founded on the Journal of D. Bourgoing, Physician to Mary Queen of Scots* (London: A. & C. Black, 1895), 247.

112. See Anthonis Mor, "Queen Mary of England, Second Wife of Philip II," Museo Prado, Madrid, P02108, and "Sir Henry Lee," National Portrait Gallery, London, NPG 2095. Pomanders themselves suggest similarly gendered practices. For long chain examples, see "Pomander Oval Body," A642076; "Silver Pomander in the Form of a Book," A641827; and "Pomander, Spherical, Embossed with Flowers," A642046, all at the Museum of Science and Industry, London. For a short chain example, see "Brass Filigree Pomander, Heart-Shaped," A35764.

113. Crooke, *Mikrokosmographia,* 223.

114. Ambroise Paré, *The Workes of that Famous Chirurgion* (London, 1630), 944.

115. Wombs represented both the uterus and the stomach. On the relationship between wombs and the uterus, see Paster, *The Body Embarrassed,* 45–46, and Schoenfeldt, *Bodies and Selves,* 13.

116. Laurent Joubert, *Erreurs Populaires* [1579], trans. Gregory David de Rocher (Tuscaloosa: University of Alabama Press, 1989), cited in, Paster, *The Body Embarrassed,* 43. Dock leaf is a common wild herb known for its healing properties in skin ailments, especially rashes from pine needles. See "Country Cures," Natural History Museum, London, www.nhm.ac.uk/nature-online/life/plants-fungi/country-cures/ (cited January 26, 2010).

117. Paré, *Workes,* 945.

118. Ibid.

119. David Cressy describes how virginity could be linked to continence: "Single people were supposed to remain continent, but married couples were supposed to make love"; *Birth, Marriage, Death: Ritual, Religion, and the Life-Cycle in Tudor and Stuart England* (New York: Oxford University Press, 1997), 290.

120. See Valerie Traub, *Renaissance of Lesbianism in Early Modern England* (New York: Cambridge University Press, 2002), chap. 2.

121. Paré, *Workes,* 945.

122. When read against other literary allusions in the period, including Ben Jonson's 1603 invocation of a "perfumed dildoe," this point is clearer; *The Workes of Benjamin Jonson* (London, 1616), sig. Z2 r.

123. *Dogs* and *monkeys* were common terms in the period for sexually promiscuous women. See Drayton's "The Owle" for the sparrow's phallic connotations. Further, Ian Moulton suggests that Jonson's jokes about these "pets" refer to artificial penises, noting that "dildoes were often seen as part of the intimate furnishings of the bedchambers of fashionable ladies (as were, say, odd pets like monkeys)." The word *dildo* seems to have originated as a "nonsense" ballad refrain—"dildo de dildo de dildo" being analogous to "fa-la-la" or "hey nonny, nonny," a generic reference to women's vaginas. For more on early modern pornography, see Ian F. Moulton, *Before Pornography: Erotic Writing in Early Modern England* (New York: Oxford University Press, 2000), 183.

124. "How then shall his smug wench, / How shall her bawd (fit time) assist her quench / Her sanguine heat? Linceus, canst thou sent? / She hath her monkey and her instrument / Smooth fram'd at Vitrio. O grievous misery!/Luscus hath left his female luxury . . . his Ganimede, / His perfum'd shee-goate, smooth-kemb'd and / high fed"; John Marston, *The Scourge of Villanie: Three Bookes of Satyres* (London, 1599), sig. C5 r–v. I am grateful to Melissa Jones for this reference.

125. For an analysis of how these dogs cite gendered cultural practices, see Ian MacInnes, "Mastiffs and Spaniels: Gender and Nation in the English Dog," *Textual Practice* 17.1 (2003): 21–40.

126. *Love-apples* was also a term for the flowers of the tomato plant; A. Marsh, *The Ten Pleasures of Marriage . . .* (London, 1682), 6–7.

127. John Marston, *The Malcontent* (London, 1604), sig. B r.

128. Ambergis, storax, galingale, musk, sweet almonds, rocket, and especially pansies were all thought to be fragrant aphrodisiacs. See Martin Levey, "Mediaeval Arabic Bookmaking and Its Relation to Early Chemistry and Pharmacology," *Transactions of the American Philosophical Society* 52.4 (1962): 1–79, 47; Dr. Schwediawer and Joseph Banks, "An Account of Ambergrise," *Philosophical Transactions of the Royal Society of London* 73 (1783): 15, 16; and Helen Bancroft, "Herbs, Herbals, Herbalists," *The Scientific Monthly* 35.3 (1932): 15.

129. N. H., *The Ladies Dictionary . . .* (London, 1694), 63–64. See also "An Electuary for the Passions of the Heart," in W. M., *The Queen's Closet,* 98.

130. Richard Levin argues that more than other women, Bianca "should beware mother," claiming that it is her mother-in-law who knowingly enables her uncle to rape her, while she schemes with Livia over a game of chess; "If Women Should Beware Women, Bianca Should Beware Mother," *Studies in English Literature* 37.2 (1997): 371–89. Others disagree with Levin's claim that Leantio's mother knowingly prostitutes Bianca: "[W]hether or not one accepts Levin's argument that the mother *deliberately* prostitutes Bianca, her negligence makes her at least the unwitting instrument of her daughter-in-law's downfall, and Bianca certainly considers her culpable"; Jennifer Panek, "The Mother as Bawd in the *Revenger's Tragedy* and *A Mad World, My Masters,*" *Studies in English Literature* 43.2 (2003): 434 n.3.

131. All citations are from Thomas Middleton, *Women Beware Women*, in *The Oxford Middleton*, ed. Gary Taylor et al. (New York: Oxford University Press, 2008).

132. See Nathaniel Richards, "Upon the Tragedy of My Familiar Acquaintance, Tho. Middleton," in Thomas Middleton, *Two New Playes* . . . (London, 1657), sig. A4 r.

133. The Duke's rape of Bianca has been critically contested; until recently, scholars have routinely described it as a seduction scene. For a discussion of the play's critical reception and an early feminist revision of the play's sexual economies, see Anthony B. Dawson, "*Women Beware Women* and the Economy of Rape," *Studies in English Literature* 27.2 (1987): 303–20.

134. Valerie Traub argues, for example, that desire in the play is based on a series of "psychic and material exchanges"; *Renaissance of Lesbianism*, 68.

135. Jean Howard, "Crossdressing, the Theatre, and Gender Struggles in Early Modern England," in *Crossing the Stage: Controversies on Cross-Dressing*, ed. Lesley Ferris (New York: Taylor & Francis), 19–50, 31.

136. All citations from *A Midsummer Night's Dream* are from Greenblatt et al., *Norton Shakespeare*, 839–896.

137. Paster, *The Body Embarrassed*, 142.

138. Tanya Pollard, *Drugs and Theatre in Early Modern England* (Oxford: Oxford University Press, 2005), 3.

139. Ibid., 1.

140. *Circulation* has been a key term in early modern scholarship. See Stephen Greenblatt, *Shakespearean Negotiations: The Circulation of Social Energy in Renaissance England* (Oxford: Clarendon, 1988), and Valerie Traub, *Desire and Anxiety: Circulations of Sexuality in Shakespearean Drama* (New York: Routledge, 1992), 5. I employ the term here to argue that discursive circulation of eroticism in the early modern period has a material counterpart: the circulation of scented materials in early modern air.

Chapter 5 · Oiled in Ambergris

1. See Phyllis Mack, *Visionary Women: Ecstatic Prophecy in Seventeenth-Century England* (Berkeley & Los Angeles: University of California Press, 1994), 15.

2. Lady Eleanor Davies, *Ezekiel the Prophet Explained as Follows* (London, 1647), 6.

3. Ibid., 5–6.

4. Lady Eleanor Davies, *Prophetic Writings of Lady Eleanor Davies*, ed. Esther Cope (New York: Oxford University Press, 1995).

5. Historian Phyllis Mack argues that although Davies's contemporaries undoubtedly interpreted her prophetic visions through religious allegory, modern historians have not, choosing instead to read it as a "form of emotional catharsis" or as a symptom of "psychic instability"; Mack, *Visionary Women*, 88. Diane Watt, however, reads this episode, and the angel's scent, allegorically: it is an annunciation scene. The angel is Gabriel and his glove "signifies that Davies is a bride of Christ"; *Secretaries of God: Women Prophets in Late Medieval and Early Modern England* (Rochester, NY: Boydell & Brewer, 2001), 153.

6. Valerie Cumming, *Gloves* (London: B. T. Batsford, 1982); Edmund Launert, *Parfüm und Flakons: Kostbare Gafässe für erlesenen Duft* (Munich: Callwey, 1985).

7. Elizabethan sampler, Metropolitan Museum of Art, New York, Object no. M-4778.

8. His titular doeskin gloves were popular in Nuremberg at the time. Joseph Leo Ko-erner, *The Moment of Self-Portraiture in German Renaissance Art* (Chicago: University of Chicago Press, 1997), 37–39.

9. Michael Snodin and John Styles, *Design and the Decorative Arts: Britain, 1500–1900* (London: Victoria & Albert Museum, 2001), 117.

10. Ann Rosalind Jones and Peter Stallybrass, "Fetishizing the Glove in Renaissance Europe," *Critical Inquiry* 28.1 (2001): 114–32.

11. See David Cressy, *Birth, Marriage, and Death: Ritual, Religion, and the Life-Cycle in Tudor and Stuart England* (New York: Oxford University Press, 1997), 263–65, 452–54.

12. *Loves Garland: or Posies for Rings, Hand-kerchers, & Gloves: And Such Pretty Tokens That Lovers Send Their Loves* (London, 1648), n.p.

13. Robert Herrick, *Hesperides, or the Works Both Humane & Divine of Robert Herrick, Esq.* (London, 1648).

14. Cumming, *Gloves*, 27. See also Jane Ashelford, *The Art of Dress: Clothes and Society, 1500–1914* (London: The National Trust, 1996), 60–63.

15. Phyllis Emily Cunnington and Catherine Lucas, *Costume for Births, Marriages, and Deaths* (New York: A & C Black, 1972), 67.

16. An early indication of this distinction is found in the household accounts of Sir John Nevile, of Chete, Knight, which lists costs associated with the marriage of his daughter, Elizabeth Nevile to Roger Rockley, including two pairs of gloves (one perfumed, one not). The pair of perfumed gloves cost 3 shillings more than the plain pair of gloves. See W. B. Redfern, *Royal and Historic Gloves and Shoes* [1904] (Berkeley, CA: Lacis, 2003), 5.

17. A contemporaneous Italian tale explains why Spaniards make the best scented gloves, even better than the French: "Nature alwaies helps mens defects, with somewhat of rare vertue; and that therefore she had given the Monopoly of making sweet Gloves to that Nation whose hand did stink insufferably"; Traiano Bocalini, *I ragguagli di Parnasso; or, Advertisements from Parnassus in Two Centuries* (London 1656), 406–7.

18. See John Florio, *Florio His Firste Fruites Which Yeeled Familiar Speech, Merie Prouerbes, Wittie Sentences and Golden Sayings* (London, 1578), 2–4.

19. Jonathan Hope, "Shakespeare's 'Native English,'" in *A Companion to Shakespeare,* ed. David Scott Kastan (Oxford: Blackwell, 1999), 239–55, 239.

20. Michel de Montaigne, *Essays,* trans. John Florio (London, 1613), sig. Q2 r.

21. In both *Troilus and Cressida* and *Henry V,* gloves facilitate recognition and identity. For an analysis of gloves in both Chaucer's and Shakespeare's *Troilus and Cressida,* see Laura F. Hodges, "Sartorial Signs in *Troilus and Criseyde,*" *The Chaucer Review* 35.3 (2001): 223–59, and Carol Chillington Rutter, *Enter the Body: Women and Representation on Shakespeare's Stage* (New York: Routledge, 2001), chap. 4. For an analysis of the role of leather-working in *Henry V,* see Alison Chapman, "Whose Saint Crispin's Day Is It?: Shoemaking, Holiday Making, and the Politics of Memory in Early Modern England," *Renaissance Quarterly* 54.4 (2001): 1467–94, 1487. In Middleton and Rowley's *The Changeling,* Beatrice and DeFlore's exchange of gloves introduces the pair's complicated sexual dynamics: "Now I know / She had rather wear my pelt tann'd in a pair / Of dancing pumps than I should thrust my fingers / Into her sockets here"; *The Changeling,* ed. Joost

Dalder (New York: A. & C. Black, 1990), 1.1.232–35. Quotations from *The New Inn* and *The Alchemist*, are taken from *The Works of Beniamin Ionson* (London, 1616).

22. Lewes Roberts, *The Merchants Mappe of Commerce* (London, 1638), 42.

23. Many goods were sold by traveling merchants, some of whom were women. Herb women were present in early modern markets; however, they did not threaten established guilds until after the Restoration. Their presence in print culture increased after 1670, as physicians began to fight with apothecarists to preserve their business. In the many pamphlets published by both sides, "herb-woman" was a common insult attacking professionalism. See Christopher Merret, *The Accomplisht Physician, The Honest Apothecary, and the Skilful Chyrurgeon* (London, 1670), 18. Although N. de Larmessin's engravings from late-seventeenth-century France are outside the scope of this project, they raise provocative questions about how perfume alters conceptions of market spaces. His engravings of both perfumers and apothecarists visually remind scholars working on market spaces that the effect of traveling vendors could be great. The perfumer's costume, for example, is composed of aromatics and includes a smoking censer, fans, and a variety of pastilles, powders ,and gloves. Similarly, his engraving of an apothecarist is made up of aromatic pills, powders, and spices; however, his apothecarist includes a rather large, hanging, phallic vial mapping the difference between the two professions through gender. See Annette Green and Linday Dyett, *Secrets of Aromatic Jewelry* (Paris: Flammarion, 1998), 62.

24. S. William Beck, *Gloves, Their Annals and Associations: A Chapter of Trade and Social History* (Detroit: Singing Tree Press, 1969), 11.

25. Herman Melville, *Moby-Dick*, ed. Tony Tanner (New York: Oxford University Press, 1998), 366.

26. Andrew Marvell, "Upon Appleton House," *Andrew Marvell: The Complete Poems*, ed. Jonathan Bates (New York: Penguin Classics, 2005), 75–99, line 180.

27. Abu-Zakariyya Yuhanna ibn Māsawayh, a ninth-century Assyrian doctor (known in the West as Johannes Damscenus of Mesue the Elder), listed five types of ambergris and their qualities in his *Treatise on Simple Aromatic Substances*. See M. Levey, "Ibn Māsawayh and His Treatise on Simple Aromatic Substances," *Journal of the History of Medicine* 16 (1961): 394–410.

28. John Parkinson, *Theatrum Botanicum: The Theater of Plants or an Universall and Compleate Herball Composed by John Parkinson, Apothecary of London, and the Kings Herbarist* (London, 1640), 1565.

29. Ibn Māsawayh first listed these five primary ingredients of perfume. See R. A. Donklin, *Dragon's Brain Perfume: A Historical Geography of Camphor* (Leiden: Brill, 1999), 25. For more on Muslim influences on Spanish perfumes, see Oliva Remie Constable, *Trade and Traders in Muslim Spain: The Commercial Realignment of the Iberian Peninsula, 900–1500* (New York: Cambridge University Press, 1999), particularly chap. 2.

30. Most fashion historians argue that the appeal of elaborate, ostentatious embroidery reached its peak during the Jacobean period and waned during the subsequent periods of civil strife that led to the English Civil War. See George Frederick Wingfield Digby, *Elizabethan Embroidery* (London: Faber & Faber, 1963), 26, and John Nevinson,

Catalogue of English Domestic Embroidery of the 16th and 17th Centuries (London: Victoria & Albert Museum, 1938).

31. See Launert, *Parfüm und Flakons,* 9.

32. See Shirley Nelson Garner, "'Let Her Paint an Inch Thick': Painted Ladies in Renaissance Drama and Society," *Renaissance Drama,* n.s. 20, ed. Mary Beth Rose (1990): 123–40, 131. Perfumers were a prominent part of Italian markets, especially in Milan and Florence, though Bianco is believed to be the first Parisian perfumer. For more on Italian luxury markets and perfumers, see Evelyn Welch, *Shopping in the Renaissance: Consumer Cultures in Italy 1400–1600* (New Haven, CT: Yale University Press, 2005), 33, 179.

33. See Cissie Fairchilds, "The Production and Marketing of Populuxe Goods in Eighteenth-Century Paris," *Consumption and the World of Goods,* ed. J. Brewer and Roy Porter (New York: Routledge, 1994), 228–48.

34. However, she left the pair on display at the Ashmolean behind. See Landsdowne MS 72, British Library, London, fol. 79. See also Renate Smollich, *Der Bisamapfel in Kunst und Wissenschaft, Quellen, und Studien Zur Geschichte Der Pharmazie* (Stuttgart: Deutscher Apotheker, 1983).

35. See Landsdowne MS 72, fol. 79.

36. Alexander Samson, *The Spanish Match: Prince Charles's Journey to Madrid, 1623* (Aldershot, U.K.: Ashgate, 2006), 37–38.

37. See Roy Strong, "Charles I's Clothes for the Years 1633 to 1635," *Costume* 14 (1980): 73–89, and Redfern, *Royal and Historic Gloves and Shoes,* plates xiv, xvii, xix, xxi, xxxi.

38. By the 1570s, perfumed gloves were routinely given as New Year's gifts, particularly by courtiers hoping to improve their favor with the in the coming year. See Lisa M. Klein, "Your Humble Handmaid: Elizabethan Gifts of Needlework," *Renaissance Quarterly* 50.2 (1997): 459–93, 463.

39. Snodin and Styles, *Design and the Decorative Arts,* 80; John Nichols, *The Progresses and Public Processions of Queen Elizabeth* (London: J. Nichols & Son, 1823), xxxix.

40. Royal MS 13, British Library, London, B. I, fol. 255.

41. Nichols, *Progresses,* 415–16.

42. See George Perfect Harding, after Nicholas Hilliard, "George Clifford, 3rd Earl of Cumberland" (ca. 1590), National Portrait Gallery, London, NPG 1492(c).

43. Anglo-Iberian trade increased throughout the sixteenth century; by the 1560s, its volume was significant. Embargos on Spanish goods were briefly issued in 1563, and from 1568 to 1573. As Pauline Croft argues, however, such embargos merely increased the difficulty of trading rather than halting practices completely; "Trading with the Enemy, 1584–1604," *The Historical Journal* 32.2 (1989): 281–302. In the 1570s volume increased tremendously, so much so that by the time the Spanish Company was granted a charter in 1577, Iberian trade was a crucial component of London's wealth. This continued until Philip II closed Spanish ports to English merchants in 1585. The Portingale community was crucial to facilitating trade between England and Spain under new political conditions. See Alan Stewart, "Portingale Women and Politics in Late Elizabethan London," in *Women and Politics in Early Modern England, 1450–1700,* ed. James Daybell (Aldershot, U.K.: Ashgate, 2004).

44. Stowe MS 167, British Library, London, fol. 102–3.

45. This was particularly true in 1598, when Scottish Catholic informer William Semple provided Spanish authorities with a list of illicit English traders working in Seville. See Croft, "Trading with the Enemy," 288.

46. Snodin and Styles, *Design and the Decorative Arts,* 109–10.

47. For more on the impact of these industries on London's sewers, see Holly Dugan, "Coriolanus and the 'Rank-Scented Meinie': Smelling Rank in Early Modern London," in *City of Vice: London 1450–1650,* ed. Roze Hentschell and Amanda Bailey (New York: Palgrave, 2010), 139–59.

48. Ibid.

49. P. J. M. Marks, *The British Library Guide to Bookbinding* (Toronto: University of Toronto Press, 1998), 43.

50. See Cummings, *Gloves,* 26. The process of dressing the leather was known as *tawing;* when dressed with alum and salt, the leather turned white, a process known as *whittawing.*

51. Robert Ashton, *The City and the Court, 1603–1643* (New York: Cambridge University Press, 1979), 72.

52. L. A. Clarkson, "The Organization of the English Leather Industry in the Late Sixteenth and Seventeenth Centuries,'" *The Economic Review* 13.2 (1960): 245–256, 246–48.

53. See Cummings, *Gloves,* 11–12, and Clarkson, "English Leather Industry," 246.

54. In 1591 English freebooters attempted to trade commodities with the king of Jerusalem for ambergris. See Andrew Dalby, *Dangerous Tastes: The Story of Spices* (Berkeley & Los Angeles: University of California Press, 2002), 66.

55. Edward IV's 1493 ban on imported goods made it illegal to import gloves; however, this outdated legislation was not easily enforced. The cessation of trade with Spain in the late sixteenth century made such imports difficult. On Spanish leather and shoemaking, see Amanda Bailey, *Flaunting: Style and the Subversive Male Body* (Toronto: University of Toronto Press, 2007), 32.

56. William Waterson and John Hill Burton, *A Cyclopædia of Commerce, Mercantile Law, Finance, Commercial Geography, and Navigation* (New York: H. G. Bohn, 1847), 343.

57. The Spanish Company primarily traded fish from the new world and corn to the Iberian Peninsula. It was finally disbanded in the House of Common's attacks on monopoly guilds in 1604. See Pauline Croft, "Free Trade and the House of Commons, 1605–06," *Economic History Review* 28.1 (1975): 17–27, 18.

58. See Clarkson, "English Leather Industry," 246.

59. Grenoble, Paris, and Randers had large gloving industries in Europe; Irish glovers, particularly in Limerick and Dublin, were known for "chicken gloves," made from unborn kid and calf skins, which were reportedly so fine they could fit into a walnut.

60. The consumption of meat increased greatly in London across the sixteenth century, due to the city's exponential population growth and the influence of Reformation restructuring of Catholic fast days.

61. Joseph Ward, *Metropolitan Communities: Trade Guilds, Identity, and Change in Early Modern London* (Stanford, CA: Stanford University Press, 1997), 3.

62. Ian Archer, "The London Lobbies in the Later Sixteenth Century," *The Historical Journal* 31.1 (1988): 17–44.

63. See MS 11588/1, Guildhall Library, London, fol. 593. This practice continued in the seventeenth century. See Guildhall Library MS 80216, 14 Charles I, part 25, no. 2, "Charter granted by King Charles to the Company of Glovers of the city of London."

64. Thomas Smith, "Discourse of the Common Weal of this Realm of England," cited in J. Thirsk, *Economic Policy and Projects: The Development of a Consumer Society in Early Modern England* (New York: Oxford University Press, 1978), 14.

65. William Stafford, *A Briefe Concepite of English Pollieye* (London, 1581).

66. George Unwin, *The Gilds and Companies of London* (London: Methuen, 1908), 257. For a reading of Unwin's account of Darcy's petition, see Garrett Sullivan, *The Drama of Landscape: Land, Property and Social Relations on the Early Modern Stage* (Stanford, CA: Stanford University Press, 1998), 220–21.

67. See Guildhall Library MS 80216. See also Ashton, *City and the Court*, 72.

68. The charter for the incorporation of the Worshipful Society of Glovers first cites the growth in numbers of those in the industry–including women–and the growth in their dependents as a primary reason for incorporation:

> "[W]hereas by an humble petition presented lately unto us by our loveing subjects liveing in and about our cittyes of London and Westminster useing the trade arte or misterie of glovers wee have been enformed that theire famylies are about fower hundred in nomber and upon them depending above three thousand of our subjects that worke upon the same arte whoe are much decayed and impoverished by reason of the great confluence of persons of the same trade arte or misterie unto our said cittyes from all parts of our kingdome of England and dominion of Wales that for the most part have scarcely served any tyme therunto working of gloves in chambers and corners and taken apprentices under them *many in number as well women* as that become burthensome to the places wherein they inhabite and are disordered multitude liveing without government and *making naughty and deceiptfull gloves* and that our subjects aforesaid that lawfully and honestyly use the said trade arte or mistery are not onely by this meanes prejudiced here at home but the reputation that Englishe gloves had in foraigne parts where they were a greate comodity and in good esteeme has much impaired (Guildhall Library MS 80216)."

69. Certain spices had to be processed, however, before they could be consumed. Garbling sandalwood, for example, was huge mercantile undertaking of the grocers' guild. Improper garbling was yet another potential source to produce "unwholesome" goods; Grocer's Company, *A Proftiable and Necessary Discourse, for the Meeting of with the Bad Garbelling of Spices* (London, 1592), sig. B4 r.

70. Linda Levy Peck, *Consuming Splendor: Society and Culture in Seventeenth-Century England* (New York: Cambridge University Press, 2005), 13.

71. Apothecarists were also members of the grocers' guilds in Leicester, Canterbury, and Salisbury. See S. W. F. Holloway, "The Regulation of the Supply of Drugs in Britain before 1868," in *Drugs and Narcotics in History*, ed. Roy Porter and Mikuláŝ Teich (New York: Cambridge University Press, 1997), 77–96, 79.

72. The physicians gained a second victory in 1563, when Mary awarded them the right to destroy the faulty goods without prior approval from the grocers should they prove uncooperative.

73. Ward, *Metropolitan Communities*, 116.

74. Lawrence Stone, "Elizabethan Overseas Trade," *Economic History Review*, 2nd ser., 2 (1949–50): 30–58, 39.

75. See Charles Raymond Booth Barrett, *The History of the Society of Apothecaries of London* (London: E. Stock, 1905), 33.

76. For records of such destructions, see Guildhall Library MS 11588/1, fol. 593, "Grocers' Company Orders of the Court of Assistants"; Guildhall Library MS 8252, "Draft of Bill Proceedings Court of Star against Grocers on Trade of Defective Medicines"; Guildhall Library MS 8293, "Draft Ordinances, 1618–1743," fol. 1, 3, 4, 5, 6, 7, 9, 16, 24.

77. *Treacle* was a generic term of art in the period, referring to any compound herbal medicine offered in a thick, sweet base; it often contained dozens of ingredients, including ginger, cinnamon, cassia, malabathrum, galbanum, cardamom, nard, pepper, frankincense, myrrh, and saffron. See Guildhall Library MS 8200, "Society of Apothecarists Court Minute Books," vol. 1.

78. See Guildhall Library MS 11588/1, fol. 593, "Grocers' Company Orders of the Court of Assistants." Stramony, stramonium, or thorn-apples are known for their narcotic properties.

79. Guildhall Library MS 8200, "Society of Apothecarists Court Minute Books," fol. 47, 56, 60.

80. Warden records from both the grocers' guild and the apothecarists' guild reflect many instances in which this occurred. See Guildhall Library MS 8252, 8293/1, and MS 8292.

81. Guildhall Library MS 8251, fol. 30.

82. Ibid., fol. 25.

83. Latin was associated with the practice of medicine. For example, William Turner, Dean of Wells Cathedral, argued that knowledge would flow down from the physicians, through the apothecarists, to herb-wives: if the system was flawed it was because physicians "commit not their knowledge of herbes unto the potecaries . . . potecaries do to the old wyves, that gather herbes, and to the grocers"; *A New Herball* (Part 1) (London, 1551), sig. A3 v. For more on the role of Latin in English medical education, see Andrew Wear, *Knowledge and Practice in English Medicine, 1550–1680* (New York: Cambridge University Press, 2000).

84. Guildhall Library MS 8251, fol. 29–31.

85. The herbal underwent significant changes in the late sixteenth century. The central herbal of the sixteenth century was Dioscorides's *De Medica Materia*, with over 75 editions published between 1501 and 1600. The Dioscoridean herbal contained four major traits: each entry describes one plant, native to the region where the book was designed to be used, and these entries are organized alphabetically by their common or vernacular names. See Jerry Stannard, *Herbs and Herbalism in the Middle Ages and Renaissance*, ed. Katherine E. Stannard and Richard Kay (Aldershot, U.K.: Ashgate Variorum, 1999), 217.

86. Parkinson, *Theatrum Botanicum*, 1, 1564.

87. Ibid.

88. For more on the influence of foreign drugs on English bodies, see Jonathan Gil Harris, *Sick Economies: Drama, Mercantilism, and Disease in Shakespeare's England* (Philadelphia: University of Pennsylvania Press, 2004), 121–22. See also Mary Floyd-Wilson, *English Ethnicity and Race in Early Modern Drama* (New York: Cambridge University Press, 2006), 56.

89. John Hemmingway, royal apothecarist to Elizabeth I, recorded large amounts of perfumes in his six-month "midsummer" inventory of 1564. See Leslie G. Matthews, *Royal Apothecaries* (London: Wellcome Historical Library, 1967), 71.

90. Peck, *Consuming Splendor*, 13.

91. As Linda Levy Peck has argued, new patents for luxury manufacturing increased substantially between 1617 and 1640; ibid., 19, chap. 2.

92. William Shakespeare, *Henry IV, Part 1*, in *The Norton Shakespeare: Based on the Oxford Edition*, 2nd ed., ed. Stephen Greenblatt, Walter Cohen, Jean E. Howard, and Katharine Eisaman Maus (New York: W.W. Norton, 2008), 1177–1254.

93. William Shakespeare, *Much Ado About Nothing*, in Greenblatt et al., *Norton Shakespeare*, 1407–70.

94. George Ford, a "Scotchman" and perfumer, resided in Aldgate as a "seruaunte, perfumer" of Lawrence Shryve, cobbler and denizen of London, along with Ford's own servant, Peter Barton.

95. These restrictions on strangers trading in London included higher taxes, restricted housing, restricted trade with other strangers, and a limit on trading periods (usually six weeks after landing). See Irene Scouloudi, *Returns of Strangers in the Metropolis, 1593, 1627, 1635, 1639* (London: Huguenot Society of London, 1985), chap. 4.

96. Charles Littleton, "Social Interactions of Aliens in Late Elizabethan London: Evidence from the 1593 Return and the French Church Consistory 'Actes,' " *Proceedings of the Huguenot Society of Great Britain and Ireland* 26.2 (1995): 151.

97. James Jacobs and Vincent Lardin were Italian perfumers and lodgers of George Wickham in the Parish of St. Clement Danes. John Nibbin, a French perfumer, resided with Mistress Cooley in the Parish of St. Martin's-in-the-Fields. The 1635 return lists "Mounsier Beaulte," a French perfumer, who resided in the Parish of St. Clement Danes in Westminster, along with William Dungan, a perfumer and housekeeper residing in the Parish of St. Martin's-in-the-Fields. Dungan, along with his Welsh partner, Thomas Richards, leased two perfume shops in the New Exchange. "Monsier" Brakan, and a "Monsier Deboe," both listed as French perfumers in the 1638 return, resided in the same parish as Dungan. Finally, the 1638 return listed a Bryan Allen, an Irish lodger in Newington town, was a "seller of sweet powder." Scouloudi, *Returns of Strangers*, chap. 4.

98. William Piddock, perfumer, submitted a bill for twelve pairs of them at four shillings for a dozen pairs. David R. Ransome, "Wives for Virginia, 1621," in *The William and Mary Quarterly* 48.1 (1991): 3–18, 16.

99. I am grateful to Thomas Dungan, who shared his research on these perfumers. See *Calendar of the Middlesex Session Records*, ed. William Hardy (London: C.W. Radcliffe, 1936), 3:32, 155, 389.

100. Ann Saunders, "Organization of the Exchange," in *The Royal Exchange*, ed. Ann Saunders (London: London Topographical Society, 1997), 89.

101. See Robert Greene, *A Disputation Between a Hee Conny-Cather, and a Shee Conny-Catcher, Where a Theefe or a Whore, Is Most Hurtfull in Cousonage, to the Common Wealth* (London, 1592), sig B2.

102. William Shakespeare, *The Winter's Tale,* in Greenblatt et al., *Norton Shakespeare,* 2881–2961.

103. Kay Staniland, "Thomas Deane's Shop in the Royal Exchange," in Saunders, *Royal Exchange,* 59–67.

104. Linda Woodbridge, "The Peddler and the Pawn: Why Did Tudor England Consider Peddlers to Be Rogues?" in *Rogues and Early Modern English Culture,* ed. Craig Dionne and Steve Mentz (Ann Arbor: University of Michigan Press, 2004), 143–171, 157.

105. George Abbot, *An Exposition upon the Prophet Ionah Contained in Certaine Sermons, Preached in S. Maries Church in Oxford* (London, 1600), 239.

106. See Annette Green, *Secrets of Aromatic Jewelry* (Paris: Flammarion, 1998), 62.

107. J. F. Merrit, *The Social World of Early Modern Westminster: Abbey, Court, and Community* (Manchester, U.K.: University of Manchester Press, 2005), 159.

108. See Jean Howard, *Theater of a City: The Places of London Comedy* (Philadelphia: University of Pennsylvania Press, 2007), 156.

109. See Hatfield House Archives, Cecil Papers, Box R5, 1633 leases, Hatfield, U.K. See also Peck, *Consuming Splendor,* 52.

110. Dungen had two other partners in his perfuming business: Thomas Richards, a Welshman, and Thomas Mazzine, a second-generation Italian.

111. See Hatfield House Archives, Cecil Papers, Box R5, 1633 leases.

112. See Folger Shakespeare Library, MS Add 901, Washington, D.C. Bolsworth was also listed as a haberdasher in a later defeasance of mortgage. See East Sussex Records Office, MS AMS6270/68, Lewes, U.K.

113. *Indian perfume* could refer to a type of Spanish tobacco, which was popular in the period. This tobacco was mixed with perfume; there was also a tobacco of "Malta," which was also mixed with aromatics. See Fernand Braudel, *The Structures of Everyday Life: The Limits of the Possible* (London: Collins, 1981), 262.

114. Halliwell Phillips MS 1239, Chetham Library, Manchester. Trunkett's perfume inventory from 1732 includes perfumed powders and pomatum.

115. Parkinson, *Theatrum Botanicum,* 1564.

116. For more on women's consumption of new kinds of ingredients and luxury goods, see Natasha Korda, *Shakespeare's Domestic Economies: Gender and Property in Early Modern England* (Philadelphia: University of Pennsylvania Press, 2002), and Wendy Wall, *Staging Domesticity: Household Work and English Identity in Early Modern Drama* (New York: Cambridge University Press, 2002).

117. As Kim Hall notes, by including recipes that called for imported goods, like sugar and spices, domestic manuals reshaped England's domestic economic policies by increasing women's appetites for imported culinary ingredients; "Culinary Spaces, Colonial Spaces: The Gendering of Sugar in the Seventeenth Century," in *Feminist Readings of Early Modern Culture: Emerging Subjects,* ed. Valerie Traub, Lindsay M. Kaplan, and Denise Albanese (Cambridge, U.K.: Cambridge University Press, 1996), 168–90, 186.

118. "The emergent topos of the housewife as keeper, as we have seen, became an apt metaphor in the early modern period to describe the wife's various, often contradictory, duties with respect to the new market goods that were infiltrating the home"; Korda, *Shakespeare's Domestic Economies,* 192. Frances Dolan paints a different picture, arguing that the home was a brutal zone and a "locus of conflict, an arena in which the most fundamental ideas about social order, identity, and intimacy were contested"; *Dangerous Familiars,* 1.

119. Hugh Plat, *Delights for Ladies to Adorne their Persons, Tables, Closets, and Distillatories* (London, 1609). In her analysis of Plat's epistle, Kim Hall argues that such imagery "while seemingly looking inward from the English seas to the English home, actually links his English woman reader to a broader colonial context"; "Culinary Spaces, Colonial Spaces," in *Feminist Readings of Early Modern Culture* (New York: Cambridge University Press, 1996), 176.

120. Plat, *Delights for Ladies,* sig. G8.

121. Another titled "King Henry the eighth his perfume" called for six spoonfuls of "Rose-water, a quarter of an ounce fine Sugar, two graines of Musk, two grains of Ambergreece, two of Civet: boile it softly together: all the house will smell of Cloves"; Plat, *Delights for Ladies,* sig. G9 v.

122. See Wall, *Staging Domesticity,* 26; Gervase Markham, *Countrey Contentments, or the English Huswife . . .* (London, 1623), 137.

123. The portion of the manuscript in Yelverton's hand begins by explaining that many of the "receipts" stem from "An Advise" from the "Queene Maides express commandement" for the "good Rules and easy Medicines without charge to the meaner sorte of People"; Wellcome Library MS 182, fol. 186.

124. Ibid.

125. It directs one to "[t]ake the buttons of Roses dryed and watered with Rosewater three or foure times put to them Muske powder of cloves Sinamon and a little mace mingle the roses and them together and putt them in little baggs of Linnen with Powder"; ibid.

126. Mrs. Hughes, "Receits Her Whole Book, Written in the Year 1637," Wellcome Library MS 363.

127. See, for example, Additional MS 28320l, Additional MS 28310, Additional MS 27466, Egerton MS 2114, Egerton MS 2214, and Egerton MS 2197, British Library; Osborn Shelves, MS B226, Beinecke Library, Yale University; and Wellcome Library MS 363.

128. Additional MS 27466, fol. 34.

129. Wellcome Library MS 182, fol. 186.

130. Ibid.

131. Simon Barbe, *The French Perfumer* (London, 1697), sig. A2 v.

Chapter 6 · Bowers of Bliss

1. *The Workes of Abraham Cowley,* ed. Thomas Sprat (London, 1668), sig. P v–P1 r. This desire echoes one expressed in an earlier Cowley poem, "The Wish," which was published in his early collection of metaphysical love poetry *Mistresse* (London, 1647), sig. C

v. In it, the poetic narrator hopes that "ere [he] descend to th' grave / May [he] a small House, and Large Garden have!"

2. Cowley's preface and poem were included in the 1666 and 1699 edition of John Evelyn, *Kalendarium Hortense, or, the Gard'ner's Almanac* (London, 1699), sig. B r. In his introductory materials, also dedicated to Cowley, Evelyn emphasizes Cowley's botanical interest and desires through the visual pleasure of gardens, juxtaposed against the artificial luxuriousness of things that pervert such beauty (undoubtedly picking up on the poem's framing of perfume). Evelyn writes: "And the Sun in his Garden gives him all he desires, and all that he would enjoy: the purity of visible Objects, and of pure Nature, before she was vitiated by Imposture or Luxury!" (sig. A4 r).

3. Cowley references Genesis 13:10. See *Workes of Abraham Cowley*, sig. P v.

4. Ibid.

5. See "Of Solitude," in ibid., sig. M3 v.

6. The stench of the Fleet sewer was legendary. Given the exponential population growth the city experienced in the late sixteenth and early seventeenth century, its sewage system was greatly overtaxed. After the great fire of 1666, Charles II passed a reform bill to provide a greater "liberty of air." See Act 19 Charles II., c. 2, section 24, cited in William Harbutt Dawson, *The Unearned Increment: or, Reaping without Sowing* (London: Swan, Sonnenschein & Co., 1890), 135.

7. Ben Jonson, "On the Famous Voyage," in *Ben Jonson and the Cavalier Poets*, ed. Hugh Maclean (New York: W. W. Norton, 1974), 14–20, lines 63–65.

8. Andrew Marvell, "The Garden," in *Miscellaneous Poems* (London, 1681), 50.

9. Evelyn, *Kalendarium Hortense*, sig. B3 r.

10. Ibid., sig. B4 v–sig. B5 r.

11. Ibid., sig. B4 v.

12. Ibid., sig. B5 r.

13. Mandy Aftel, *Essence and Alchemy: A Natural History of Perfume* (New York: North Point Press, 2001), 111.

14. Common jasmine, *Jasminum officinale,* also known as "poet's jasmine," has small white flowers and is native to India, though it was cultivated in England since the mid-sixteenth century; its famous derivation *Jasminum grandiflorum* was cultivated in Grasse, France, in the seventeenth century. It is still a key ingredient used in commercial perfumers (including Chanel No. 5). Red jasmine, or frangipane, derives from a Jamaican shrub *Plumeria rubra.* Its name stems from a fifteenth-century Italian nobleman, Mario Muzio Frangipani, who created a perfume from this flower to scent gloves; by the mid-sixteenth century, *frangipane* was used to describe scented almond pastry. Frangipane's odd etymological mobility through fashion, food, and perfume indicates the pervasive use of perfume in a variety of applications as new ingredients were introduced into English consumer culture, yet it also marks scent as a way to retain an "exotic" quality. Despite its inclusion in a popular satiric depiction of ladies' closets in Restoration drama and in a translation of the Turkish language of flowers used to indicate love (both published in the 1690s), frangipane is rarely included in most manuscript or published smoke perfume recipes in the seventeenth century, perhaps due to its rare occurrence and prohibitive cost.

15. See Alain Corbin, *The Foul and the Fragrant: Odor and the French Social Imagination* (Cambridge, MA: Harvard University Press, 1986), 38.

16. See Alan Macfarlane, *The Culture of Capitalism* (London: Basil Blackwell, 1987); Roy Strong, *The Renaissance Garden in Europe* (London: Thames & Hudson, 1979); and Keith Thomas, *Man and the Natural World: Changing Attitudes in England, 1500–1800* (London: Penguin, 1984). More recently, scholars have revisited the significance of gardening to early modern cultural studies. See Rebecca Bushnell, *Green Desire: Imagining Early Modern English Gardens* (Ithaca, NY: Cornell University Press, 2003).

17. Medieval pleasure gardens were organized by utilitarian and emblematic purposes. Although these garden spaces also designed, cultivated, and celebrated many pleasant scents, English medieval pleasure gardens were large spaces on royal or monastic grounds. See Bushnell, *Green Desire*, 9.

18. Evelyn, *Kalendarium Hortense*, sig. B4 v.

19. Raymond Williams, *The Country and the City* (New York: Oxford University Press, 1973), 32. The erasure of labor in country house discourses serve as a cultural fantasy about property and citizenship in the period, what Kari McBride terms a "script, set, and cast" for performances of legitimacy; Kari Boyd McBride, *Country House Discourse in Early Modern England* (Aldershot, U.K.: Ashgate, 2001), 3.

20. Peter Martin, *Pursuing Innocent Pleasure: The Gardening World of Alexander Pope* (Hamden, CT: Archon, 1984), xv. I differ in this assessment: the cultural significance of gardens, like perfumes, cannot be understood merely by re-creating their material components. See the introduction for an analysis of why methodologies like this one depend on implicit assumptions about the history of the human body.

21. Dianne Harris, "The Postmodernization of Landscape: A Critical Historiography," *Journal of the Society of Architectural Historians* 58.3 (1999): 434–44.

22. Most scholars cite Alexander Pope's Twickenham as the *de facto* example of the English style. In 1719, Alexander Pope took up residence at Twickenham and began planning his stately garden. His documents allow scholars to reconstruct the material history of eighteenth-century gardens, often with gendered consequences. See Susan Groag Bell, "Women Create Gardens in Male Landscapes: A Revisionist Approach to Eighteenth-Century English Garden History," *Feminist Studies* 16.3 (1990): 471–91. Anglo-Dutch gardens were popular in the late eighteenth century; see John Dixon Hunt and Erik de Jong, eds., *Journal of Garden History* 8 (1988). Renaissance gardens were greatly influenced by the Italian style; John Dixon Hunt, ed., *The Italian Garden: Art, Design, and Culture* (Cambridge, U.K.: Cambridge University Press, 1996). In the late seventeenth century, many elements of the "English" style were imported from the French. For an analysis of the development of French garden styles, see William Howard Adams, *The French Garden, 1500–1800* (New York: George Braziller, 1979), and Chandra Mukerji, "Reading and Writing with Nature: Social Claims and the French Formal Garden," *Theory and Society* 19.6 (1990): 651–79.

23. For an in-depth discussion of eighteenth-century gardening techniques, especially in Pope's Twickenham garden, see Michael Leslie and Timothy Raylor, eds., *Culture and Cultivation in Early Modern England: Writing and the Land* (Leicester, U.K.: Leicester University Press, 1992), and Martin, *Pursuing Innocent Pleasures*, 39.

24. Increased traffic in plants was linked to increased traffic in slaves, providing English merchants with a way to increase their profits on return journeys home from eastern slave markets in Istanbul or to plantations in the new world. Plants like "aromatic" sedge, for example, arrived in England from such circumnavigations, connecting odoriferous pleasure in English gardens with broader patterns of consumption of luxury goods. See Amy Tigner, "Flowers of Paradise: Botanical Trade in Sixteenth- and Seventeenth-Century England," *Global Traffic: Discourses and Practices of Trade in English Language and Culture from 1550 to 1700*, ed. Barbara Sebek and Stephen Deng (Houndsmill, U.K.: Palgrave Macmillan, 2008), 137–56, 144.

25. See Benedict Robinson, "Green Seraglios: Tulips, Turbans, and the Global Market," *Journal for Early Modern Cultural Studies* 9.1 (2009): 93–122, 97.

26. Ibid., 107.

27. See Bushnell, *Green Desire*, chap. 1.

28. Alain Corbin argues the opposite: "[A]ny attempt to train the sense of smell always results in disappointment. This is why olfaction would not be taken into account in designing the English garden as the privileged place of sensory education and fulfillment"; *Foul and the Fragrant*, 7. Scholars of English gardens disagree with this assessment. For discussions of principles of medieval garden design, see Ramona Jablonski, *The Medieval Garden Design Book* (Owings Mills, MD: Stemmer House, 1982), and Sylvia Landsberg, *The Medieval Garden* (New York: Thames & Hudson, 1996). This new style echoes changes in the print history of gardening manuals. Rebecca Bushnell summarizes this shift: "The earlier books let the gardener-readers delight in finding their own way through the book and the garden; the reader was the master of the book, just as he or she was meant to manage the disorder of nature. The later books, however present readers with an ordered garden and book, rather than trusting them on their own"; *Green Desire*, 51.

29. For a greater discussion of the evolution of labor in relationship to the field of gardening, particularly applied to Hyll's *Profitable Arte of Gardening*, see Bushnell, *Green Desire*, 41–43.

30. Thomas Hyll, *The Profitable Arte of Gardening* (London, 1568), sig C v. See also Bushnell, *Green Desire*, 93.

31. Strong, *Renaissance Garden*, 165.

32. For more on Eden and early English gardens, see Tigner, "The Flowers of Paradise."

33. See, for example, a seventeenth-century example of grafting: Pietro de Crescenzi, *De Omnibus Agruculturæ Partibus Animaliuma* (Basel, 1548).

34. See also "How to make Apples, Pears, and other Fruit of several colours, and to give them a dainty tast of Spices" in John White, *A Rich Cabinet with Variety of Inventions in Several Arts and Sciences* (London, 1677), 3:52. For a contemporaneous discussion of experiments that attempt to alter a plant's color or scent, see John Parkinson, *Paradisi in Sole Paradisus Terrestris: A Garden of All Sorts of Pleasant Flowers Which Our English Ayre Will Permitt* (London, 1629). See also Anonymous, *The Expert Gardener . . .* (London, 1654), and *The Orchard and the Garden* (London, 1594). By 1690, "grafting" began to expand beyond its botanical application. See Anonymous, *Reasons Humbly Offered against Grafting . . .* (London, 1690).

35. See Arthur W. Hill, "The History and Functions of Botanic Gardens," *Annals of the Missouri Botanical Garden* 2.1–2 (1915): 185–240.

36. By the late seventeenth century, it was obvious that no garden could display the world's entire natural splendor, fueling a desire to develop universal taxonomies that would occupy eighteenth-century scientific endeavors. As Richard Drayton has written, it "forced a second loss of Eden"; *Nature's Government: Science, Imperial Britain, and the "Improvement" of the World* (New Haven, CT: Yale University Press, 2000), 19.

37. For an analysis of England's horticultural spaces as constructing national identity in the seventeenth century, see Amy Tigner, "'England's Paradise': Horticultural Land-scapes in the Renaissance" (Ph.D. diss., Stanford University, 2004). For the eighteenth century, see Timothy Morton, *The Poetics of Spice: Romantic Consumerism and the Exotic* (New York: Cambridge University Press, 2000).

38. See Bushnell, *Green Desire*, 52, 70, 57.

39. See Joan Thirsk, "Making a Fresh Start: Sixteenth-Century Agriculture and the Classical Inspiration," in Leslie and Raylor, *Culture and Cultivation*, 16–34, 20.

40. John Worlidge, *Systema Horti-Culturæ, or, the Art of Gardening in Three Books . . .* (London, 1677).

41. See, for example, Walter Charleton, *The Immortality of the Human Soul, Demonstrated by the Light of Nature* (London 1657), 30.

42. See Robert Boyle, *An Experimental Discourse of Some Unheeded Causes of the Insalubrity and Salubrity of the Air* (London, 1690); Francis Bacon, *Sylua Syluarum: or a Naturall Historie in Ten Centuries* (London, 1627), 218–20; and John Harington, *The History of Polindor and Flostella, with Other Poems* (London, 1657), 94–95.

43. William Coles, *The Art of Simpling, An Introduction to the Knowledge and Gathering of Plants* (London, 1656), 92.

44. Ibid.

45. There are several recipes for purifying the air of pestilence through smoke perfumes or through burning scented candles. In recipe books of the 1630s, most however involve vinegars and would create a very different scent. See "To make Fumes and Perfumes. An excellent perfume to cast a sent in a chamber and against the ill aire," in Philbert Guibert, *The Charitable Physitian . . .*, trans. I. W. (London, 1639), sig. F r.

46. Strong, *Renaissance Garden*, 118.

47. Ibid., 19.

48. Ibid., 45.

49. Ralph Austen, *A Treatise of Fruit Trees . . .* (Oxford, 1665), 58–59.

50. Austen also images that, in some cases, gardens exhale artificially perfumed air. On attempts to graft new scents or colors onto roses; see ibid. See also Parkinson, *Paradisi in Sole Paradisus Terrestris*, 25.

51. Henry Hawkins and Herbert Aston, *Partheneia Sacra . . .* (Rouen, 1633), 17, 36, 44, 48.

52. Ibid., 5. For more on Hawkins's text see Christine Coch, "In a Lady's Bower: Poetry, Gardens, and the Problem of Pleasure in Early Modern England" (Ph.D. diss., University of Chicago, 2002).

53. Anonymous, *The Orchard and the Garden* (London, 1594), sig. G2 r–G3 v.

54. See Worlidge, *Systema Horti-Culturæ*, 6.

55. Ibid.

56. The earliest imported porcelain contained many varieties of floral patterns. See John Ayers, O. R. Impey, and J.V.G. Mallet, *Porcelain for Palaces: The Fashion for Japan in Europe, 1650–1750* (London: Oriental Ceramic Society, 1990), and David Battie, *Sotheby's Concise Encyclopedia of Porcelain* (Boston: Little, Brown, 1990).

57. See Corbin, *Foul and the Fragrant*, 77–80. On French pleasure gardens, see Chandra Mukerji, "Reading and Writing with Nature: Social Claims and the French Formal Garden," *Theory and Society* 19 (1990): 651–79. The earlier the English vase is, the larger its size. As potpourri vases increased in popularity, their size and price decreased.

58. In the late sixteenth century, dining ware, teapots, and drinking cups represented the bulk of imported porcelain; over the course of the next fifty years, new porcelain objects entered Europe's markets. Mass-marketed chinoiserie was not available in England prior to 1744, due to several factors. First, Chinese porcelain manufacturers closely guarded the recipe for the translucent white porcelain that fascinated Europeans. A number of factories in England attempted to recreate this white mixture using ingredients such as lead, ox-bones, or flint, to no avail. The royal factories of the French and Dutch were finally able to recreate this mixture in the early eighteenth century. Worker espionage eventually made the porcelain recipes from Meissen and Sèvres more widely available in Europe. See "The Spread of Trade Secrets" in Hilary Young, *English Porcelain, 1745–1795: Its Makers, Design, Marketing, and Consumption* (London: Victoria & Albert Museum, 1999), 54–73. Second, unlike the royal porcelain factories of France and Holland, England's porcelain factories were purely commercial. The most notable example was the Chelsea porcelain factory, although by the late eighteenth century, at least thirteen commercial English factories were in existence. See Appendix 1, in Young, *English Porcelain*, 197–99.

59. English ships departing from European ports used flint as ballast, storing it in the hold. Vessels might carry up to a hundred tons, which could be sold for as much as £10 a ton—a massive profit margin. Returning ships needed to find a commodity that could replace the flint as ballast. See Battie, *Sotheby's Concise Encyclopedia of Porcelain*, 50. On England's sixteenth-century and early-seventeenth-century trading expeditions to the Far East, see Richmond Barbour, *Before Orientalism: London's Theatre of the East, 1576–1626* (Cambridge, U.K.: Cambridge University Press, 2003), 40.

60. Young, *English Porcelain*, 181.

61. Geoffrey A. Godden, *Oriental Export Market: Porcelain and Its Influence on European Wares* (London: Granada, 1979), 42.

62. Young, *English Porcelain*, 7.

63. See David Porter, *Ideographia: The Chinese Cipher in Early Modern Europe* (Stanford, CA: Stanford University Press, 2001). See also Karen Newman, "City Talk: Women and Commodification in Jonson's *Epicœne*," *English Literary History* 56.3 (1989): 503–18.

64. Lorna Weatherill, *Consumer Behaviour and Material Culture in Britain, 1660–1760* (London: Routledge, 1988), 31. Weatherill's study relies on orphans court inventories; it is possible that wives' purchases were recorded under husbands' names. See Young,

English Porcelain, 190. On women's porcelain collecting, see also B. Kowaleski-Wallace, "Women, China, and Consumer Culture in Eighteenth-Century England," *Eighteenth-Century Studies* 29.2 (1995–96): 153–67.

65. J. H. Plumb, *In the Light of History* (London: Houghton Mifflin, 1972), 59.

66. Porter, *Ideographia,* 135.

67. Ibid.

68. Corbin identifies the increasing relationship between foul and fragrant smells with social networks of power, arguing that there is a progression away from the use of strong, animal-based scents toward lighter floral scents (or in the present framework, from Spanish perfumes to French ones). Those who smelled strongest were clearly marked as dirty, poor, immoral, unfashionable, and rural, and thus needed the strongest perfumes to cover noxious bodily odors. Corbin locates the shift in taste as part of a broader movement toward Enlightenment scientific discourse and the development of a middle-class bourgeoisie. As the discourse of medicine increasingly moved away from Galenic models of health, the strong association between powerful musks, sexuality, and excrement-made animal-based perfumes too overt for polite society; Corbin, *Foul and the Fragrant,* 68, 81.

69. See Nicholas Culpeper et al., *Pharmacopœia Londinensis; or, the London Dispensatory* (London, 1654), 44; Vignau Du and John Phillips, *The Turkish Secretary . . .* (London, 1688); and Mary and John Evelyn, *Mundus Muliebris; or, the Ladies Dressing-Room Unlock'd . . .* (London, 1690).

70. The link between odoriferous garden pleasures and those exhaled from pot-pourri vases strengthens in the eighteenth century in Chelsea, England. Home to England's first porcelain factory, Chelsea was known as a unique site of scented pleasure prior to its opening in 1741. Proximate the fashionable pleasure gardens of Ranelagh, Vauxhall, and to Sir Henry Sloanes's newly designed Worshipful Society of Apothecaries' Garden, or more commonly known today as the Chelsea Physic Garden, the Chelsea porcelain factory emerged alongside of the floral markets of this area of London. The two histories are linked, so much so that Chelsea porcelain factory owner Nicholas Sprimont would commission a porcelain series advertised as "enameled from Sir Hans Sloane's Flowers"; this series, produced from 1752 to 1758, was based on illustrations of Apothecaries' Garden specimens in the *Figures of the Most Beautiful, Useful and Uncommon Plants* (1752), published by the guild's head gardener, Philip Miller. The series was so popular that it was soon copied in factories in Bow, Derby, and even on imported porcelain from the Far East. See Sue Minter, *The Apothecaries' Garden: A New History of Chelsea Physic Garden* (Stroud, U.K.: Sutton, 2000).

71. Sir Kenelm Digby, *Choice and Experimented Receipts in Physick and Chirurgery* (London, 1688), 299–302; Gervase Markham, *Countrey Contentments, or the English Huswife . . .* (London, 1683), 109; Hugo Plat, *Delights for Ladies . . .* (London, 1617); W. M., *A Queens Delight . . .* (London, 1654), 262, 74. Manuscript recipes include "A Perfume to Burn" (probably added later by a second hand, in roundhand) in British Library MS 27466, London. See also Francis Evans's 1696 unpublished cookbook, Osborne 238b, Beinecke Library, Yale University, New Haven, CT. See also "A Receipt to Make Pastills to Burn,"

"Rose Pastills," "Pastills of Roses Made into Wax Tapers," "Spanish Pastills," "Portugal Pastills," "A Receipt to Make Perfumed Paste for Beads and Medals," "Another Sort," "Another Sort," "Another Sort" in Simon Barbe, *The French Perfumer* ... (London, 1697), 75, 79.

72. Potpourri vases are on display at Versailles, the Metropolitan Museum of Art, the Philadelphia Art Museum, the Museum of Costume History in Bath, and the Victoria and Albert Museum.

73. See Anonymous, "Covered Jar, Mounted in Ormolu as Pot-Pourri and Fountain," and "Fish, mounted in European Ormolu as a Fountain a Parfum," both late seventeenth century, Musée Royaux d'Art et d'Histoire, Brussels.

74. For an analysis of the ways pleasure gardens function within women's poetry in the period, see the introduction to Coch, *In a Lady's Bower.* For an analysis of the garden as an extension of women's domestic space, see Catherine Alexander, "The Garden as Occasional Domestic Space," *Signs* 27.3 (2002): 857–71.

75. See MacDougall, *Medieval Gardens.*

76. See Dympna Callaghan, "(Un)Natural Loving: Swine, Pets, Flowers in *Venus and Adonis,*" *Early Modern Culture: An Online Seminar* 3 (2003): 31.

77. The strong association between gardens of pleasure and Eden situates sexual sin in Paradise's splendor. Religious pleasure and (sinful) sexual pleasure are linked in the sixteenth century, emerging from the same ontological garden. See, for example, Thomas Becon, *Fortresse of the Fayethfull* (London, 1550), and Heinrich Bullinger, *The Golden Boke of Christen Matrimonye,* trans. Miles Coverdale (London, 1542). During the first quarter of the seventeenth century, however, references to Eden as a garden of pleasure increasingly are used in travel literature as a religious mooring in distant geographies. Pleasure gardens in Egypt and Assyria are linked to Eden, signaling to Christian audiences both the abundance of providence and the exotic seduction of sin. See John Cartwright, *The Preachers Travels Wherein Is Set Downe a True Journall to the Confines of the East Indies* ... (London, 1611).

78. Barbe, *The French Perfumer,* sig. A3 r.

79. All citations from "Samson Agonistes" are from John Milton, *Critical Edition of the Major Works,* ed. Stephen Orgel and Jonathan Goldberg (New York: Oxford, 1991), 671–716, line 720–22, 711–12.

80. Edmund Spenser, *The Faerie Queene* (London, 1596), 385, 2.12.

81. Stephen Greenblatt, *Renaissance Self-Fashioning* (Chicago: University of Chicago Press, 1980), 170.

82. Ibid., 177.

83. For a critique of Greenblatt's invocation of temperance, see Michael Schoenfeldt, *Bodies and Selves in Early Modern England: Physiology and Inwardness in Spenser, Shakespeare, Herbert, and Milton* (Cambridge, U.K.: Cambridge University Press, 1999), 52.

84. Ibid.

85. Ibid., 50–51.

86. "Likewise that same third Fort, that is the *Smell* / Of that third troupe was cruelly assayd: / Whose hideous shapes were like to feends of hell, / Some like to hounds, some

like to Apes, dismayd, / Some like to Puttockes, all in plumes arayd: / All shap't according their conditions, / For by those vgly formes weren pourtrayd, / Foolish delights and fond abusions, / Which do that sence besiege with light illusions" (2.6).

87. See Schoenfeldt, *Bodies and Selves,* 67.

88. All citations from *Paradise Lost* are from Milton, *Critical Edition of the Major Works,* 355–618.

89. See Timothy Morton, *The Poetics of Spice: Romantic Consumerism and the Exotic* (New York: Cambridge University Press, 2006), 70–71.

90. Asmodeus's fishy fumes advance the relationships between sensual air and sexual pleasure. In the apocryphal book of Tobitt, Sarah burnt fish intestines as incense in her home in order to to repel the demon Asmodeus, who tortured her by killing each of her seven husbands before the marriage could be consummated. Milton's metaphoric link between Satan's sniffing of Paradise and Asmodeus's inhalation of fishy fumes inverts the standard role of scents in pleasure gardens; here, sweet smells offend rather than refresh Satan's spirits. Asmodeus's "fishy fumes" perhaps inadvertently alludes to another perfuming ingredient mined from the South Seas during sailing and slaving expeditions to the East: ambergris.

91. "To all delight of human sense exposed / In narrow room nature's whole wealth, yea more, / A heaven on earth, for blissful Paradise / Of God the garden was, by him in the east / Of Eden planted" (9.209–10).

92. For a discussion of the smell of London's sewers, see Holly Dugan, *"Coriolanus* and the Rank-scented *Meinie:* Smelling Rank in Early Modern London," in *Masculinity and Metropolis of Vice,* ed. Amanda Bailey and Roze Hentschell (Basingstoke, U.K.: Palgrave, 2010), 139–59.

93. Women were also linked with porcelain collecting: Jacobean drama, for example, contains references to English women's predilection for china. Ben Jonson's Mrs. Otter, in *Epicœne or the Silent Woman,* made her fortune through the china trade; *Epicœne,* ed. R. V. Holdsworth (London: A. C. Black, 1979), 3.1. See also Karen Newman, *Fashioning Femininity and English Renaissance Drama* (Chicago: University of Chicago Press, 1991), 133. Numerous representations of crazed women foolishly collecting porcelain appear in Restoration English drama; by 1710 this image was a standard stock figure for the English satirists. For a discussion of such eighteenth-century representations, see Kowaleski-Wallace, "Women, China and Consumer Culture."

94. See Elizabeth Sauer and Lisa M. Smith, *"Noli Me Tangere:* Colonialist Imperitives and Enclosure Acts in Early Modern England," in *Sensible Flesh: On Touch in Early Modern England,* ed. Elizabeth Harvey (Philadelphia: University of Pennsylvania Press, 2003), 141–58.

95. Thomas, *Man and the Natural World,* 220.

96. Strong begins his treatise on royal gardens in England in 1660. See *Royal Gardens* (New York: Pocket Books, 1992), 7.

97. Elisabeth MacDougal, *Fountains, Statues, and Flowers: Studies in Italian Gardens of the Sixteenth and Seventeenth Centuries* (Washington, D.C.: Dumbarton Oaks Research Library & Collection, 1994).

98. Ibid., 200.

99. All citations from "The Convent of Pleasure" are from Margaret Cavendish, Duchess of Newcastle, *The Convent of Pleasure and Other Plays,* ed. Anne Shaver (Baltimore: Johns Hopkins University Press, 1999), 217–47.

100. Lady Happy catalogs the following fabrics: "Silk-Damask," "Taffety," "Gilt Leather," "Tapestry," "fine Mats," "Turkie Carpets," "Silk quilts," "fine Holland," "fine Diaper and Damask," and "finest and purest Linen." She orders service on "plate of Purseline," "Gilt-plate," and orders only "choisest Meats every Season doth afford, and every day our Meat, be drest several ways" (2.2).

101. For an exploration of "convents," see Theodora Jankowski, *Pure Resistance: Queer Virginity in Early Modern English Drama* (Philadelphia: University of Pennsylvania Press, 2000).

102. Misty Anderson, "Tactile Places: Materializing Desire in Margaret Cavendish and Jane Barker," *Textual Practice* 13.2 (1999): 329–52, 333.

103. Ibid.

104. Ibid.

105. Ibid. Ann Shaver, in her introduction to the modern critical edition of the *Convent of Pleasure And Other Plays,* agrees, but she adds that they also mimic Elizabethan approaches to the classical unities; Cavendish, *Convent of Pleasure,* 1–17, 10–11.

106. Anderson, "Tactile Places," 332.

107. Strong, *Renaissance Garden,* 1–15.

108. Nature aligns with Lady Happy's beauty, Fortune, her wealth, and Heaven her virtuousness. For a reading of the queerness of Lady Happy's virginity, see Jankowski, *Pure Resistance,* 177–78.

109. As Karen Raber points out, Lady Happy abandons her utopian project of distributing pleasure to those who are without, inviting only "happy," i.e., noble women like herself (with the gifts of nature, fortune and heaven); *Dramatic Difference: Gender, Class, and Genre in Early Modern Closet Drama* (Newark: University of Delaware Press, 2001).

110. The invocation of death here suggests erotic pleasure. Interestingly, this surfeit of pleasure in the senses does present some issues of health. In a throwaway line that describes the interior of the convent to two men, Mr. Take-Pleasure and his servant, Dick, Madam Mediator mentions that the convent has "Women-Physicians, Surgeons and Apothecaries" as well as the Lady herself, Votress to Nature, serving as "Confessor" (2.1).

111. Anderson, "Tactile Places," 355.

112. See Valerie Traub's reading of this scene in *The Renaissance of Lesbianism in Early Modern England* (Cambridge, U.K.: Cambridge University Press, 2002), 178–79.

113. Ambergris is a product of the male sperm whale. For more on its early modern meanings, see chapter 5.

114. Laura Rosenthal, *Playwrights and Plagiarists in Early Modern England: Gender, Authorship, Literary Property* (Ithaca, NY: Cornell University Press, 1996), 91, cited in Traub, *Renaissance of Lesbianism,* 291.

115. Traub, *Renaissance of Lesbianism,* 291.

116. Ibid.

117. Ibid.

118. Given the number of classical literary allusions to sexuality, in both pastoral and lyric traditions in this period, it is hard to argue that they are not. This pleasure, however, is gendered. See Bruce R. Smith, *Homosexual Desire in Shakespeare's England: A Cultural Poetics* (Chicago: University of Chicago Press, 1995), 82. See also David Halperin, *Before Pastoral: Theocritus and the Ancient Tradition of Bucolic Poetry* (New Haven, CT: Yale University Press, 1983).

119. Anthony Fletcher, *Gender, Sex, and Subordination in England, 1500–1800* (New Haven, CT: Yale University Press, 1999), 53; Angus McLaren, *Reproductive Rituals: The Perception of Fertility in England from the Sixteenth to the Nineteenth Century* (New York: Metheun, 1984), 14, both cited in Traub, *Renaissance of Lesbianism*, 81.

120. See Michel Foucault, *The History of Sexuality: An Introduction, Vol. 1* (New York: Vintage, 1990), 31.

121. England's importation of porcelain, of which potpourri vases made up only a portion, increased dramatically across the seventeenth century. Japanese porcelain in this period differs from Chinese porcelain in several key ways: first, the Japanese ports were closed to all European merchants but the Dutch after 1640. Although the English had attempted to institute a porcelain "factory" in Japan, they abandoned the plan in 1623. The decreased supply of Japanese porcelain led to the second major difference between Chinese and Japanese porcelain: its incredible expense. Due to its high value and rarity, England's aristocrats and royals mostly collected Japanese porcelain. I am more interested in popular understandings and uses of porcelain and therefore focus instead on Chinese porcelain. The large collection of Japanese porcelain in England's museums, however, demonstrates the fluidity of exportation and re-exportation in the period. Such pieces could have been procured through Dutch merchants or through Chinese merchants in India. For a greater discussion of Japanese porcelain, see Ayers, Impey, and Mallet, *Porcelain for Palaces*.

Conclusion · Ephemeral Remains

1. Carl von Linné and Stephen Freer, *Linnaeus' Philosophia Botanica* (New York: Oxford University Press, 2003), 45. See also Richard Grove, *Green Imperialism: Colonial Expansion, Tropical Island Edens, and the Origins of Environmentalism, 1600–1800* (Cambridge, U.K.: Cambridge University Press, 1996), 13.

2. See George D. Ehret's illustration of Linnaeus's system, which first appeared in the *Species Planatarum* (Leiden, 1736), cited in Mary Louise Pratt, *Imperial Eyes: Travel Writing and Transculturation* (New York: Routledge, 1992), 25.

3. See Linné and Freer, *Linnaeus' Philosophia Botanica*, 231.

4. Ibid.

5. For more about how the narrative of the Enlightenment shapes its intellectual history and contemporaneous cultural meaning, see Daniel Brewer, *The Enlightenment Past: Reconstructing Eighteenth-Century French Thought* (New York: Cambridge University Press, 2008), esp. 1–23.

6. See, for example, Alexander Pope, *Dunciad* (London, 1729), Book 3, line 217.

7. Denis Diderot, "Letter on the Blind," *Oeuvres complètes,* 25 vols. (Paris: Hermann, 1978), 4:27. For more on Diderot's approach to the blind, see Nicholas Mirzoeff, "Bindness and Art," *The Disability Studies Reader,* 2nd ed., ed. Leonard Davis (New York: Routledge, 2006), 382–95.

8. See Susan Scott Parrish, *American Curiosity: Cultures of Natural History in the Colonial British Atlantic World* (Chapel Hill: University of North Carolina Press, 2006), 258.

9. Ibid.

10. Laurinda S. Dixon, *Perilous Chastity: Women and Illness in Pre-Enlightenment Art and Medicine* (Ithaca, NY: Cornell University Press, 1995), 31.

11. Several ancient writers differentiated between *sagicitas,* or acumen, and *sapientia,* or philosophical wisdom.

12. Ben Jonson, *A New Inn* (The Revels Plays), ed. Michael Hattaway (Manchester, U.K.: Manchester University Press, 2001). I am thankful to Barbara Sebek for alerting me to this exchange.

13. See Edward Hall, *The Union of the Two Noble and Illustre Famelies of Lancastre and Yorke* (London, 1548), sig. BBB1 r.

14. Ibid., sig. CCC1 v.

15. William Drummond, "Quinque Sensus," Folger MS ADD 1246.24, Folger Shakespeare Library, Washington, DC, quoted in Jonathan Gil Harris, "The Smell of *Macbeth,*" *Shakespeare Quarterly* 58.4 (2007): 465–86, 477. Harris's article discusses in fascinating detail King James I's nasal prowess and its relationship to the early modern stage.

16. Lancelot Addison, *The Life and Death of Mahumed* (London, 1679), sig. B2 v.

17. Dryden and Pope both use the term to describe bloodhounds; *Oxford English Dictionary,* 2nd ed., s.v. "sagacious," no. 1.

18. Max Horkheimer and Theodor W. Adorno, *Dialectic of Enlightenment: Philosophical Fragments* (Palo Alto, CA: Stanford University Press, 2002), 151.

19. Ibid.

20. As many critics have argued, such temporal alignment moves certain bodies into modernity while rendering others as primitive, pungent, and backward.

21. To sum up: "*Ambrosial* things are restorative, *fragrant* things are exciting, *spicy* things are stimulating, *noisome* things are stupefying, and *nauseous* things are corrosive." In 1756, Linneaus amended this list to include two more: *odores hircine,* or lascivious smells (based on the smells of a male goat), and *odores alliacei,* or garlicky smells. Linné and Freer, *Linnaeus' Philosophia Botanica,* 103.

22. Michel Serres, *The Five Senses: A Philosophy of Mingled Bodies* (London: Continuum, 1985), 169.

23. Ibid., 170.

24. See Frances Bacon, "*Catalogvs Historiarvm Particvlarivm,*" appended to *Francisci de Verulamio, Summi Angliae Cancellarii, Instauratio magna* (London, 1620), sig. D1 v, cited in Carla Mazzio, "The History of Air: *Hamlet* and the Trouble with Instruments," *Harboiled* 26.1–2 (2009): 153–96. Mazzio's insightful argument informs my analysis here.

25. John Donne, *Devotions upon Emergent Occasions* (New York: Vintage, 1999), 79, cited in Mazzio, "History of Air," 153.

26. Robert Boyle, *An Experimental Discourse of Some Unheeded Causes of the Insalu-brity and Salubrity of the Air* (London, 1690), sig. D3 v.

27. See Corbin, *Foul and Fragrant*, 225.

28. Phthalates are used as a solvent in most mass-marketed perfumes; the problem is compounded in mass-marketed perfumes scented with musk.

29. Serres, *Five Senses*, 179.